Bioenergy Crops for Ecosystem Health and Sustainability

The growing of crops for bioenergy has been subject to much recent criticism, as taking away land which could be used for food production or biodiversity conservation. This book challenges some commonly-held ideas about biofuels, bioenergy and energy cropping, particularly that energy crops pose an inherent threat to ecosystems, which must be mitigated.

The book recognises that certain energy crops (e.g. oil palm for biodiesel) have generated sustainability concerns, but also asks the question 'is there a better way' of using energy crops to strategically enhance ecosystem functions? It draws on numerous case studies, including where energy crops have had negative outcomes as well as cases where energy crops have produced benefits for ecosystem health, such as soil and water protection from the cropping of willow and poplar in Europe and the use of mallee eucalypts to fight salinity in Western Australia. While exploring this central argument, the volume also provides a systematic overview of the socio-economic sustainability issues surrounding bioenergy.

Alex Baumber is a Postdoctoral Research Fellow and Sessional Lecturer in Interdisciplinary Environmental Studies at the University of New South Wales, Australia. He has previously worked for the Future of Australia's Threatened Ecosystems (FATE) Program and for the Australian Government Department of Environment and Heritage, Sustainable Wildlife Industries Section.

Bioenergy Crops for Ecosystem Health and Sustainability

Alex Baumber

Routledge
Taylor & Francis Group
LONDON AND NEW YORK

earthscan
from Routledge

First published 2016
by Routledge

2 Park Square, Milton Park, Abingdon, Oxfordshire OX14 4RN
52 Vanderbilt Avenue, New York, NY 10017

Routledge is an imprint of the Taylor & Francis Group, an informa business

First issued in paperback 2018

British Library Cataloguing-in-Publication Data
A catalogue record for this book is available from the British Library

Library of Congress Cataloging in Publication Data
Names: Baumber, Alex, author.
Title: Bioenergy crops for ecosystem health and sustainability / Alex
    Baumber.
Description: New York : Routledge, 2016. | Includes bibliographical
    references and index.
Identifiers: LCCN 2015037151
Subjects: LCSH: Energy crops--Environmental aspects. | Energy
    crops--Economic aspects.
Classification: LCC SB288 .B38 2016 | DDC 333.95/39--dc23
LC record available at http://lccn.loc.gov/2015037151

ISBN: 978-1-138-83883-3 (hbk)
ISBN: 978-0-367-17322-7 (pbk)

Typeset in Goudy
by HWA Text and Data Management, London

Contents

Figures

Tables

Preface

This book is the culmination of almost ten years' work exploring the role that energy cropping could potentially play in helping to restore the Earth's degraded and vulnerable landscapes. This has included a PhD thesis at the University of New South Wales (UNSW), research grants from the Australian Government and rural industry bodies, a variety of activities under the auspices of Bioenergy Australia and, most importantly, the opportunity to interview landholders and other stakeholders in the agriculture and energy industries as they work through various options for energy cropping and how they could be implemented in different contexts.

When I first began to work on bioenergy issues, I was aware of some of the controversies surrounding deforestation, dispossession and food security that certain bioenergy crops had attracted. However, I was also intrigued by the possibility that crops such as mallee eucalypts or willow could deliver a very different set of outcomes, especially in the rural landscapes where I had been working as part of the FATE (Future of Australia's Threatened Ecosystems) programme at UNSW.

Reconciling the two competing visions of bioenergy crops that I had been exposed to – destroyers of forests and food security on the one hand, and restorers of landscape resilience on the other – seemed at first to be an impossible task. However, by exploring the complexities of a wide range of sustainability issues that go well beyond local landscapes, such as greenhouse gas balances, impacts on global food supply and international policy developments around the promotion of renewable energy, I came to see that it was not only possible to devise energy cropping systems that are truly sustainable, but also that by doing so we might be able to shed new light on the very concept of sustainability itself.

Other land use activities, such as food production or forestry, have also had to grapple with complex sustainability issues, including limits to natural resource use, climate change, biodiversity loss and the protection of fundamental human rights. However, the global expansion of bioenergy cropping has brought these issues together like no other land use activity before. This book attempts to pick these issues apart while elucidating the interconnections between them. Hopefully, along the way it also helps to expand the boundaries of bioenergy sustainability as it is commonly conceptualised – by expressly considering how energy crops can help to restore and protect degraded and vulnerable landscapes.

A special thank you goes out to all those who have assisted me in refining my thinking on bioenergy issues. This includes all those who have reviewed this book and my other published work, my three PhD supervisors, John Merson, Mark Diesendorf and Peter Ampt, and of course the landholders and other participants who have so generously given their time and expertise in support of research projects over a number of years, especially in Condobolin and the NSW Central Tablelands. Lastly, I wish to thank my partner Claire for her amazing support and patience during the writing process, and my sons Simon and Leo for their total lack of patience with the writing process – thus ensuring that I took the essential breaks required to maintain sanity during an endeavour such as this.

Part 1

Introduction

Chapter 1

Bioenergy crops and sustainability

Feeding the growing demand for biodiesel is likely to take place through expanding palm oil plantations in Indonesia. Big commodity traders are already planning significant expansion in the biodiesel infrastructure. Once this is established, it will feed off forest destruction and fuel not only cars but climate change.

Greenpeace, *How the Palm Oil Industry is Cooking the Climate* (2007)

... the present agrofuels frenzy is likely to lead to an uncontrolled expansion in palm-oil production in many parts of South-east Asia. Just over half (55 per cent) of the region's peatlands remain undrained, and it seems almost inevitable that over the next few years almost all of it will be transformed into giant biodiesel plantations, mainly of oil palm. Barring a policy U-turn, this will lead to an additional 42–50 billion tonnes of carbon being belched into the atmosphere in the coming years.

A. Ernsting, Agrofuels in Asia: Fuelling *Poverty, Conflict, Deforestation* (2007)

Millions of hectares of customary forestlands are being taken to produce oil palm in Indonesia and parts of Malaysia ... In the Philippines, "idle, under-utilised lands" which are also the traditional common lands of communities, are being allocated for biofuels, food, and rubber plantations.

W. Anseeuw et al., *Land Rights and the Rush for Land* (2012)

Species diversity in oil palm plantations is much less than in natural forests, even degraded forests. Forest clearing for oil palm leads to species losses.

D. Sheil et al., *The Impacts and Opportunities of Oil Palm in Southeast Asia* (2009)

The quotes presented here represent just some of the concerns that have been voiced about the expansion of oil palm plantations in southeast Asia and the role that is being played in this by the growing global demand for biofuels. While biofuel demand is not the only reason for the expansion of oil palm plantations, and some progress has been made in breaking the link between biofuels and deforestation, oil palm for biodiesel in southeast Asia remains the archetypal example of how biofuel demand can lead to biodiversity loss, the destruction of carbon stocks, land degradation and the displacement of vulnerable communities. For some critics of biofuels, such as Greenpeace (2007) and Ernsting (2007), it is particularly ironic that the very products that have been promoted as a climate change solution through policy measures such as the European Union's Renewable Energy Directive (RED) may in fact be contributing to the release of vast amounts of stored carbon from forests and peatlands, raising the question "is the cure worse than the disease?" (Doornbosch and Steenblik, 2007).

Recent analysis of satellite imagery indicates that three of the top five countries for percentage forest loss between 2000 and 2012 were in southeast Asia (Figure 1.1). For Indonesia, this satellite imagery showed an acceleration in the rate of forest loss over this period, with a noticeable uptick in deforestation after 2010 (Hansen et al., 2013). This has occurred despite the imposition of a moratorium on forest clearing in Indonesia and the introduction of biofuel certification schemes by the European Union (EU) and the Roundtable on Sustainable Biomaterials (RSB). Indonesia's 2011 moratorium, which prevents the granting of new concessions for clearing or logging on peatlands and old-growth forest,

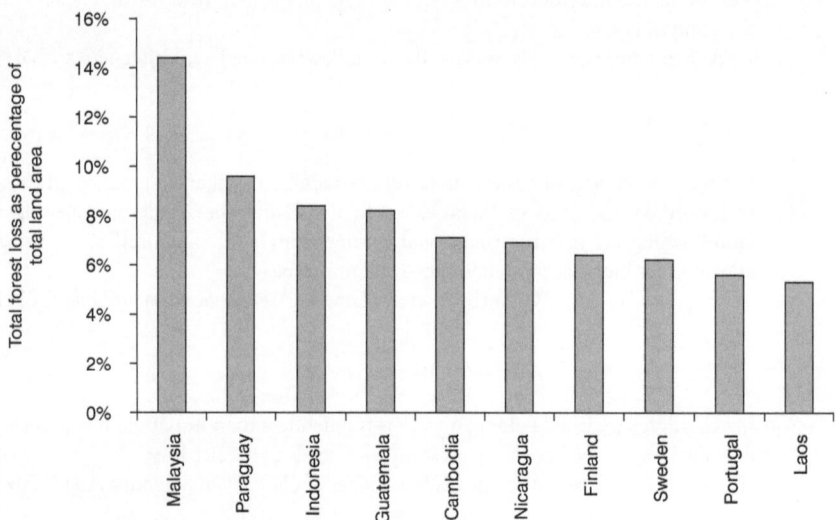

Figure 1.1 Forest loss between 2000 and 2012 as a percentage of total land area (excluding water bodies)

Source: Hansen et al. (2013)

has been criticised for classing large areas of regrowth forest as "degraded" and allowing clearing to continue in these areas (Edwards and Laurance, 2011).

While it has been estimated that around three-quarters of the palm oil produced globally is used for food and only a minority for biodiesel (Schoneveld, 2010), it is the rapid growth in biodiesel demand that has placed it at the forefront of concerns around deforestation. For example, the use of palm oil for food and other non-energy products in the European Union expanded by only 6 per cent between 2006 and 2012 (Figure 1.2). This compares to growth of 365 per cent in the use of palm oil for biodiesel over the same period. This rapid growth in biofuel demand is also apparent in a recent review of large-scale land transactions in developing countries compiled by the International Land Coalition (Anseeuw et al., 2012), which found that biofuel production was the intended purpose for over half of the recorded transactions and that Africa had become the latest frontier in this "land rush".

Based on these opening paragraphs, you could be forgiven for thinking that this book is simply another warning about the perils of biofuels, urging us to dramatically reduce demand and move away from the idea of growing bioenergy crops at all. It isn't. Rather, this book asks: is there a better way? A way that can not only avoid negative impacts such as deforestation, degradation and dispossession, but can actually help to restore or protect degraded or vulnerable landscapes? And if such a way forward can be found, how do we get there?

Bioenergy crop (or simply *energy crop*) is a broad term that can refer to a wide range of crops used to fulfil various energy needs. The crops grown for bioenergy may be common agricultural crops like corn or soy, woody perennials like oil

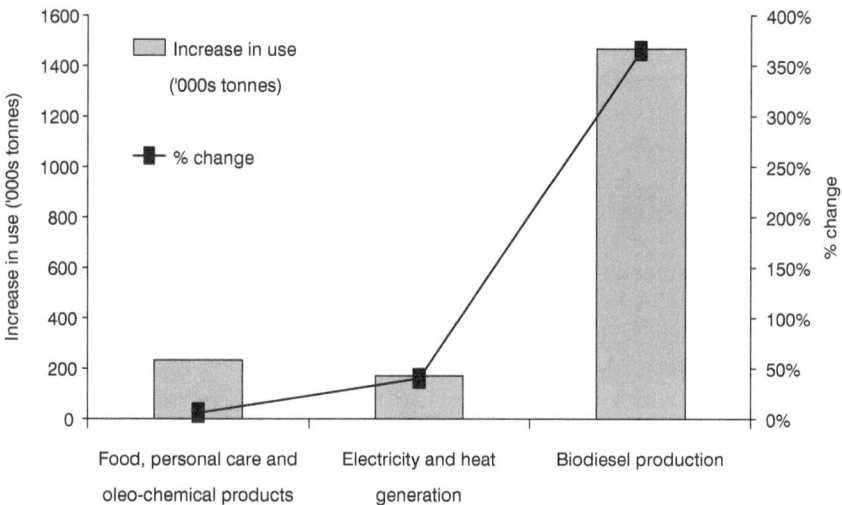

Figure 1.2 Growth in the use of palm oil in the European Union 2006–12
Source: Gerasimchuk and Koh (2013)

Figure 1.3 Willow grown as an energy crop in Europe
Source: reproduced with kind permission from Kevin Lindegaard, Crops for Energy Ltd
(www.crops4energy.co.uk)

palm or eucalyptus, or perennial grasses like miscanthus or switchgrass. The fuels produced can be solid, liquid or gaseous and may be used for a variety of purposes, including transport, electricity generation or heating. Co-production is also common, with the bioenergy component of the cropping system ranging from a minor by-product (e.g. electricity from rice husks) to the primary purpose (e.g. corn grown exclusively for ethanol production).

Given the diversity of contexts in which bioenergy crops may be grown, it is only to be expected that the impacts of energy cropping will be equally diverse. Box 1.1 highlights an example of energy cropping that is less well-known than oil palm, but has shown very different impacts to those presented at the start of this chapter.

The delivery of environmental and/or social benefits from commercial cropping systems as discussed in Box 1.1 is sometimes referred to as *multifunctionality*. A key aim of this book is to promote a broad understanding of the term "sustainable bioenergy" that includes the promotion of positive, multifunctional outcomes rather than simply focusing on ways to prevent negative impacts. To borrow a line from William McDonough and Michael Braungart, whose *cradle-to-cradle* concept advocates a revolutionary change to the way we use materials and energy, we need to shift our thinking away from how to be "less bad" towards a focus on maximising the good (McDonough and Braungart, 2002). Ensuring the sustainability of energy crops is not just about mitigating the risks that crops such as oil palm might pose to biodiversity, climate and local communities, it is also about designing and promoting forms of bioenergy that actively restore degraded

Box 1.1 **Environmental benefits from short-rotation cropping of poplar and willow**

Short-rotation cropping of poplar and willow involves high density plantings that are harvested frequently, up to once a year, with plants re-sprouting from the base through a process known as *coppicing*. These woody crops have been promoted in Europe in particular for a range of commercial products, including timber, animal fodder and bioenergy. Unlike oil palm plantations, the primary bioenergy products are electricity and heat rather than liquid biofuels – although this may change as advances are made around so-called *second-generation* biofuels that can be produced from woody biomass.

Simpson et al. (2009) highlight a number of examples from Europe and North America of short-rotation bioenergy crops contributing to local environmental enhancements, including increases in soil organic matter, improved water quality and enhanced biodiversity. Short-rotation crops have also been used successfully to filter wastewater (Schroeder, 2012) and to remove metals such as cadmium and zinc from contaminated soils (Laureysens et al., 2005). Dimitriou et al. (2011) found good evidence that short-rotation woody crops in Europe generally improve groundwater quality, reduce heavy metal concentrations in soils and increase the abundance and diversity of birds, but also cautioned that other impacts can be more variable and require further research.

Apart from Europe, short-rotation woody crops are grown in temperate countries from Canada to New Zealand. There is also increasing interest in developing countries such as India, where woody energy crops have been targeted at wastewater treatment (Riddell-Black et al., 2012) and as a strategy for both mitigating and adapting to climate change (Swamy, 2012). They have also been suggested as a means of resolving the potential conflict between food and fuel production by utilising "marginal land", although Weih (2012) cautions that care is required to avoid negative impacts on biodiversity.

land, reduce greenhouse gas emissions and create viable livelihood options for local people.

Willow and poplar are not the only energy crops with the potential to enhance ecosystem health while providing commercial returns. Other energy crops that are discussed in subsequent chapters include:

- Mallee eucalypts, which can be coppiced like willow or poplar to provide biomass for electricity generation and biofuel production while helping to mitigate salinity in Western Australia (Stucley et al., 2012);

- Switchgrass, which could provide feedstock for ethanol while reducing soil erosion, increasing soil carbon and providing habitat for birds in the Midwestern United States (Hartman et al., 2011); and
- Jatropha, a shrub with oil-rich seeds which could be used to produce biodiesel while helping to combat desertification in the Sahel region of Africa (Holthuijzen and Maximillian, 2011).

While each of these crops has shown some potential to enhance ecosystem health, it is important to recognise that these benefits are dependent on the way that cropping is undertaken and the fact that an energy cropping system has shown positive results in one context does not mean that such benefits will be universal. Similarly, a focus on maximising the good does not mean that negative impacts should be ignored. Simpson et al. (2009) point out that location, species and management practices all have an influence on the impacts from growing poplar and willow, while Bennett et al. (2011) advise that mallee trees will only reduce salinity impacts in Western Australia under specific circumstances. Furthermore, the impacts of energy cropping depend very much on what was on the land before, whether it be existing farmland, degraded land or biodiverse primary forest. Jatropha plantings may well have the potential to restore degraded land, but they can also be implicated in the clearing of existing woodlands, with Romijn (2011) finding that the "carbon debt" created by clearing Africa's miombo woodlands for jatropha would take several decades of biofuel production to repay.

The complexities determining the environmental benefits and risks of these energy cropping examples are dealt with in more detail in Chapters 2, 3 and 4, which focus on climate change, deforestation and ecological restoration, respectively.

Bioenergy basics

Before delving deeper into the various issues and examples of energy cropping covered in the subsequent chapters of this book, a few notes on terminology are needed. Some key terms have already been discussed, such as *bioenergy crop* (used interchangeably with *energy crop* in this book), which describes plants that are grown to meet human energy demands. *Energy cropping* describes the process of developing and managing an *energy cropping system*, which includes not only the land on which the crop is grown, but also the infrastructure, ecosystems, communities and markets that enable the cropping activity to occur.

The term *bioenergy* can be defined broadly to refer to energy "produced from organic matter or biomass" (UN Energy, 2007, p. 3), although a few caveats should be noted. First, the term bioenergy is generally not taken to include coal or other fossil fuels derived from plants that lived long ago, so a clearer definition is "energy from material *recently* derived from plants and animals" (Rutovitz and Passey, 2004, p. 11, emphasis added). Second, the energy obtained by humans from food crops is not generally included when using the term bioenergy, despite this being the most fundamental way in which humans utilise the energy contained in biomass. This exclusion is rarely stated but is implicit in discussions around the use of crops for food *or* bioenergy (i.e. the *food versus fuel* debate discussed in Box 1.2).

Overall, bioenergy accounts for around 10 per cent of the world's primary energy supply, more than any other form of renewable energy (Bauen et al., 2009). However, over 80 per cent of global bioenergy consumption takes the form of traditional cooking and heating fuels such as wood, charcoal, dung and agricultural residues. Less than 20 per cent of global bioenergy use qualifies as *modern bioenergy*, such as electricity, transport fuels and industrial process heat, which represent the primary focus of this book (IEA Bioenergy, 2007).

As discussed previously, a key feature of modern bioenergy is its diversity, with a wide range of feedstocks, conversion pathways and products involved (Figure 1.4). While many of the feedstocks for bioenergy are wastes or residues from agricultural, forestry or industrial processes, the focus of this book is on production systems in which bioenergy is the primary purpose (or at least a major purpose) of production. Thus, the most relevant feedstocks, processes and products are those that flow from the box marked "energy crops" in Figure 1.4.

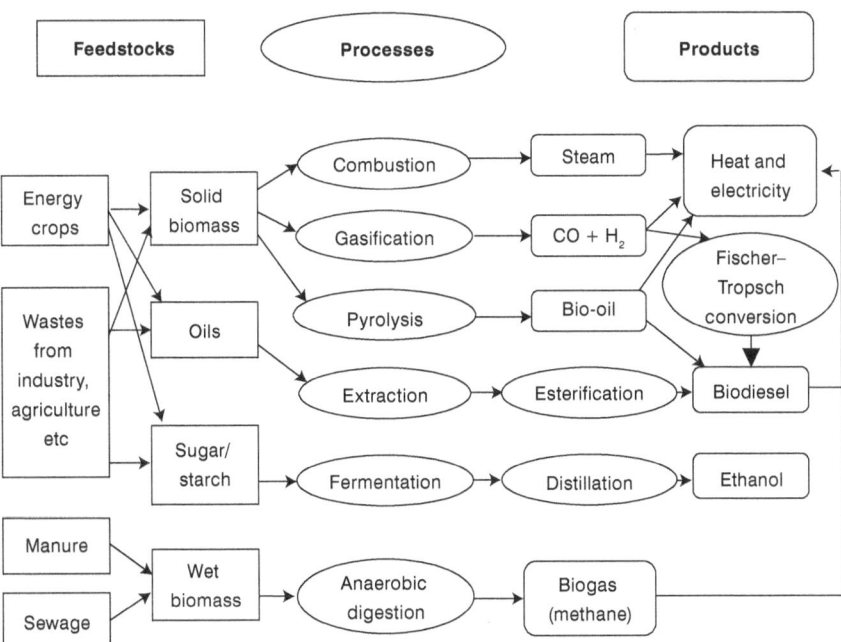

Figure 1.4 Common bioenergy feedstocks, processes and products
Source: adapted from International Energy Agency (2007)

The most common energy crops used for liquid biofuels are sugarcane and corn (maize) for ethanol, as well as oil palm, soy and rape (canola) for biodiesel. For electricity and heat, common energy crops include grasses such as miscanthus and trees such as willow or poplar (Karp and Halford, 2011). The term *biofuel* can technically be used to refer to all manner of fuels produced from biomass, whether they are solid, liquid or gas. However, it is very common when looking at media reports, policy documents and even research papers to find the term biofuel applied exclusively to liquid biofuels, such as ethanol and biodiesel. This practice has also been adopted for this book, unless otherwise noted.

While Figure 1.4 shows a clear demarcation between wastes and energy crops, in practice a grey area exists in which it is not always clear whether a particular source of biomass is simply a waste or residue of another process or whether it should be considered a co-product in its own right. A good example of this is sugarcane processing, where different grades of sugar and molasses are produced, along with bagasse (residual biomass left after extraction of cane juice) and these different components may be used for food, animal feed, ethanol, heat or electricity generation depending on market conditions and the technologies available. An industry may start out with a focus on a single product, but over time the co-production of multiple products from what were previously seen as wastes may become integral to its ongoing viability.

For the purposes of this book, bioenergy does not need to be the sole product or even the primary product of a cropping system in order for the crop to qualify as an energy crop. What is important is that the bioenergy component is of sufficient value as to influence decision-making around how and where the crop might be grown.

Sustainability basics

Having introduced some of the key terms and concepts related to energy cropping, we now turn our attention to the concept of sustainability. Depending on who you listen to, energy cropping may be cast as either a sustainability hero, leading us away from our current unsustainable reliance on fossil fuels, or a sustainability villain, driving rapid environmental destruction and social upheaval. In order to understand the two faces that energy cropping can present, it is useful to first explore what we mean by *sustainability* and *sustainable development*, terms that have been defined in hundreds of different ways since they rose to prominence in the 1970s and 1980s.

A fundamental element of sustainability is long-term persistence, reflecting concerns raised in seminal works such as *The Limits to Growth* (Meadows et al., 1972) that human activities are depleting our resource base and threatening fundamental ecological support systems. This focus on long-term persistence is reflected in the following description of sustainability by Stephen Dovers of the Fenner School of Environment and Society at the Australian National University (ANU):

> Sustainability refers to the ability of human society to persist in the long term in a manner that satisfies human development demands but without threatening the integrity of the natural world.
>
> (Dovers, 2005, p. 7)

The concept of sustainable development takes this notion of long-term persistence a step further by recognising that human society is not static and that further development is required to ensure that the needs of all people are met in an equitable way. While much has been written on sustainable development over the past quarter century, the most commonly cited and influential definition of sustainable development remains that presented in 1987 by the World Commission on Environment and Development (also known the Brundtland Commission):

> Sustainable development is development that meets the needs of the present without compromising the ability of future generations to meet their own needs.
>
> (World Commission on Environment and Development, 1987, p. 43)

Under the Brundtland definition, the goal of long-term persistence is expressed as a principle of inter-generation equity, with current generations having a responsibility to conserve the resources needed by future generations to meet their needs. This principle has been cited in arguments both for and against energy cropping. For advocates of energy cropping, our reliance on fossil fuels is the principle sustainability concern, threatening the long-term persistence of not only our resource stocks, but also of our climate through the release of greenhouse gases (e.g. European Parliament and Council of the European Union, 2009). To this end, the key benefits of bioenergy are its renewable nature and the fact that any carbon dioxide (CO_2) released during combustion is directly offset by CO_2 taken up during the plants' growth phase (sometimes referred to as being *carbon neutral*). In contrast, fossil fuels are non-renewable (at least at time scales commensurate with their current rates of extraction) and emit carbon to the atmosphere that was previously sequestered underground.

Opponents of energy cropping remind us that renewable doesn't necessarily mean sustainable, pointing to the threat that energy cropping can pose to the long-term persistence of forests and other critical ecosystems, along with the communities that depend on them (e.g. Greenpeace, 2007; Ernsting, 2007; Anseeuw et al., 2012). Furthermore, as will be discussed in Chapter 2, energy crops are rarely carbon neutral when their full life-cycle is considered (i.e. cultivation, harvest, processing, transport and combustion). The greenhouse gases (GHGs) emitted across the full life-cycle of an energy crop generally exceed the amount of CO_2 absorbed during the plants' growth and may even result in higher net emissions than those from fossil fuels, especially if the production system involves extensive land-clearing and/or use of nitrogen-based fertilisers (e.g. Crutzen et al., 2007; Gibbs et al., 2008).

Apart from the environmental impacts of bioenergy production and use, the references of the Brundtland Commission to "development" and "needs" highlight that sustainability also has social and economic dimensions. Indeed, sustainability is commonly described as having three "pillars" – the environmental, the economic and the social. When dealing with the social aspects of sustainability, the focus is not so much on long-term persistence, but rather on promoting positive change. As authors such as Peter Marcuse argue, it is equally possible to sustain an unjust society as it is to sustain a just one, and our goal must first be to make unjust societies just before it is appropriate to sustain them (Marcuse, 2006).

The point about enhancing versus maintaining social conditions is emphasised by the 2015 release of the final report on the Millennium Development Goals, which were set in the year 2000 (United Nations, 2015). This report shows that, despite significant progress in some areas such as clean drinking water, we still have a long way to go in achieving an equitable global society. Sub-Saharan Africa and South Asia in particular fell short of a number of the goals relating to factors such as extreme poverty, hunger and child mortality.

While the Brundtland definition of sustainable development is the most prominent, it is far from being universally accepted and has been subjected to various criticisms. Keiner (2006, p. 2) contends that it is so inclusive and vague that it "loses its integrity as a political concept", while Lawn (2001, p. 13) believes that it represents "establishment appropriation" of the sustainability concept. Diesendorf (2000) criticises it for allowing trade-offs between environmental and social values, instead proposing a notion of sustainable development that would require social and environmental outcomes to be simultaneously enhanced or maintained.

Applying sustainability principles to the bioenergy sector

While definitions of sustainability such as those cited from Stephen Dovers and the Bruntland Commission tend to present it as an all-encompassing concept, in practice the incorporation of sustainability principles into individual industry sectors often occurs in a piecemeal fashion. Different sustainability issues rise to prominence at different times, generally in response to activities that are seen as posing a threat to sustainability. The importance of responding to threats in order to ensure sustainability is encapsulated in the *precautionary principle*, which arose from the 1992 Rio Conference on Environment and Development and states that "where there are threats of serious or irreversible damage, lack of full scientific certainty shall not be used as a reason for postponing cost-effective measures to prevent environmental degradation" (United Nations Conference on Environment and Development, 1992, Principle 15). However, perspectives inevitably vary on the relative importance of different threats and issues and it is only by combining these various perspectives that a picture begins to emerge of what sustainability might mean in practical terms within a given industry sector such as bioenergy.

In order to highlight the different ways in which bioenergy sustainability can be conceptualised, Table 1.1 compares three different approaches to cataloguing the key sustainability issues affecting the bioenergy sector. The aim here is not to discuss each sustainability issue in detail, as they are dealt with in turn in subsequent chapters.

The first column of Table 1.1 shows the sustainability criteria for liquid biofuels under the European Union's Renewable Energy Directive (RED). The EU has decided to promote the use of biofuels due to their potential greenhouse and energy security benefits, but does not wish to stimulate the production of biofuel feedstocks that pose a direct threat to areas of high biodiversity value, peatlands or areas with high carbon stocks. As such, the eligibility criteria shown in Table 1.1 can be seen as defining those biofuels that the EU considers worthy of being promoted. The second column presents the global principles of the Roundtable on Sustainable Biomaterials (known as the Roundtable on Sustainable Biofuels until 2013), which can be used for voluntary certification of biofuels (e.g. to reassure consumers that biofuels do not have unsustainable impacts) as well as being used by government agencies to define eligibility for biofuel support programmes. The final column shows the 2007 framework from UN Energy (a United Nations agency), which is the broadest but also the least prescriptive of the three approaches, being designed to assist with general consideration of sustainability issues rather than assessment and certification of individual biofuels.

The RSB principles shown in Table 1.1 have been endorsed by the EU as a recognised voluntary scheme that covers all of the eligibility requirements of the RED. This means that a biofuel supplier could choose to demonstrate compliance with RED rules by showing evidence directly to the EU or by having their biofuels certified by the RSB (or one of the other 19 schemes recognised by the EU as of January 2015 (European Commission, 2015). However, it is also important to note that the RSB standards go further than the EU's basic eligibility rules in areas such as greenhouse gas savings (requiring 50% instead of 35%), labour rights, land rights, community development and food security. This highlights how sustainability can be defined differently by different stakeholders.

Climate change appears in the each of the three frameworks shown in Table 1.1 and is cited as a key goal behind policy measures such as mandates and subsidies that have been used to promote bioenergy use in a range of countries. Mandates are government requirements placed on fuel or electricity companies to use a certain amount of bioenergy, while subsidies are designed to give bioenergy a competitive advantage over fossil fuels and often take the form of tax breaks (more detailed analysis of bioenergy policy measures is provided in Chapters 8 and 9).

The Fifth Assessment Report of the Intergovernmental Panel on Climate Change (IPCC), released in 2014, concludes that increases in atmospheric concentrations of CO_2 and other greenhouse gasses since the industrial revolution

Table 1.1 Sustainability issues contained in three frameworks for bioenergy sustainability

EU Renewable Energy Directive (European Parliament and Council of the European Union, 2009)	RSB Global Principles v2.1 (Roundtable on Sustainable Biomaterials, 2010)	UN Energy Sustainable Bioenergy Framework (UN Energy, 2007)
Each member state shall: • Meet renewable energy targets by 2020 Adopt a renewable energy action plan Biofuel eligibility for targets: • Greenhouse gas emissions saving (35% versus fossil fuel comparator) • Raw materials shall not be obtained from land with high biodiversity value (primary forest, nature protection areas, highly biodiverse grassland) • Raw materials not obtained from land with high carbon stock (including wetland and continuously forested area) • Raw materials not obtained from peatland that has been drained European Commission to report every 2 years on food security and labour issues	1 Compliance with laws and regulations and international laws and agreements 2 Consultative impact assessment and management process and economic viability analysis 3 Climate change mitigation (50% greenhouse gas reduction versus fossil fuel baseline) 4 Human rights or labour rights 5 Contribution to social and economic development 6 Food security 7 Impacts on biodiversity, ecosystems, and conservation values 8 Soil degradation and soil health 9 Water quality and quantity (surface and groundwater) 10 Air pollution 11 Production efficiency and risk management 12 Land rights and land use rights	1 Energy services for the poor 2 Agro-industrial development and job creation 3 Health and gender 4 Implications for agriculture 5 Food security 6 Government budget implications 7 Trade, foreign exchange balances and energy security 8 Biodiversity and natural resource management 9 Climate change

are "*extremely likely* to have been the dominant cause of the observed warming since the mid-20th century" (IPCC, 2014, p. 4), with all mitigation scenarios modelled by the IPCC requiring a substantial upscaling of "low-carbon energy" as a proportion of the total primary energy supply. However, not all forms of bioenergy can be considered low-carbon energy when their full life-cycle emissions are taken into account. As such, the EU and RSB standards include criteria to ensure that only biofuels with life-cycle emissions substantially lower than those of a comparable fossil fuel are certified as sustainable (35% lower for the EU, 50% lower for RSB). A more detailed discussion of this issue is provided in Chapter 2.

The other major driver of government support for bioenergy has been energy security, which like climate change mitigation, stems from a perceived lack of sustainability around the fossil fuels that dominate global energy supplies. Energy security tends to be most prominent in relation to oil and hence has played a greater role in the promotion of liquid biofuels than for bioelectricity. The importance of energy security can vary between different contexts. For example, in the EU it is listed as the second goal of the Renewable Energy Directive after climate change (European Parliament and Council of the European Union, 2009), while in the United States, Federal Government policy documents generally list energy security (or "energy independence") ahead of climate change when listing the reasons for promoting bioenergy (e.g. Biofuels Interagency Working Group, 2010). Some authors have also argued that the true primary motivation behind the promotion of ethanol in the US has been to support the incomes of corn farmers (e.g. Rubin et al., 2008). Figure 1.5 shows the rapid growth in US ethanol production between 2000 and 2010.

Concerns around energy security can vary in scale from the local (e.g. providing modern energy services for local communities) to the national (e.g. replacing fuel imports) to the global (e.g. impacts of a peaking oil supply on the global economy). The UN Energy framework in Table 1.1 focuses mostly on local-scale energy services in developing countries, while US and EU policies focus primarily on the replacement of imported oil with domestic supplies. At the global scale, concerns surrounding an imminent peaking of global oil supplies and continuing high prices (so-called *peak oil*) tended to be more prominent around 2005–8 (e.g. Hirsch et al., 2005; Future Fuels Forum, 2008), before being diminished somewhat by recent increases in oil supplies from sources such as shale oil in the US. For example, the US Energy Information Administration's reference case for global oil prices in its 2014 International Energy Outlook has prices continuing to decline to 2025 (Energy Information Administration, 2014).

It is not just in relation to energy security that liquid biofuels have attracted more attention than other forms of bioenergy. They have also been at the heart of the so-called *food versus fuel* debate (see Box 1.2), as well as concerns around deforestation and dispossession in developing countries (e.g. Oxfam International, 2007; Eide, 2008; Brown, 2008). In particular, these concerns have arisen around what are often termed *first-generation* biofuels, which rely on common agricultural

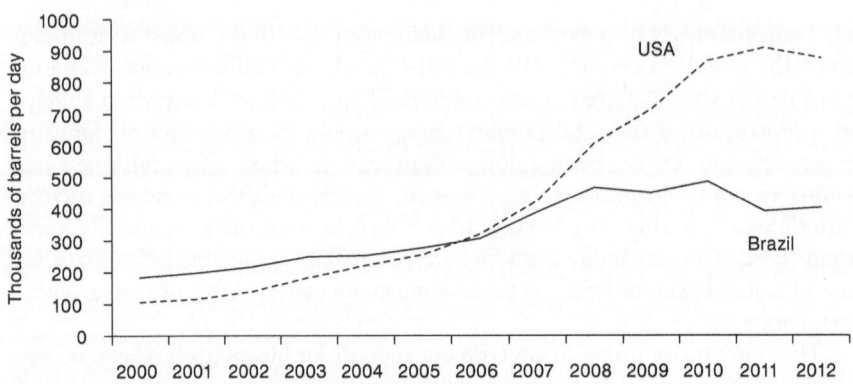

Figure 1.5 US and Brazilian ethanol production 2000–2010
Source: Energy Information Administration (2015)

crops for feedstocks and on conversion techniques that were well-established before the end of the twentieth century. These first-generation biofuels are predominantly produced from sugar or starch crops via fermentation (e.g. ethanol from sugarcane or grains) or from oil-bearing plants via transesterification (e.g. biodiesel from canola or palm oil).

The EU provides the clearest example of liquid biofuels being singled out for increased scrutiny when it comes to sustainability. The EU sustainability criteria for liquid biofuels shown in Table 1.1 are not applied to solid fuels like wood pellets, with the European Commission (2010) arguing that solid biomass fuels are less likely to be traded internationally and more likely to be produced from wastes or residues that do not cause land use change. In contrast, liquid biofuels produced from common agricultural crops are widely traded and their use can have direct impacts on global prices for food and animal feed, with flow-on effects such as providing incentives to clear land for new crops.

As highlighted in Box 1.2, one of the responses to the *food versus fuel* debate has been a strong global research and development focus on so-called *second-generation* biofuels, which can be produced from the woody or fibrous parts of plants rather than the sugar, starch or oil components. These woody and fibrous components are termed lignocellulose, as they are predominantly composed of lignin and cellulose (with the cellulose fraction being the main focus for biofuel development). For the end user, second-generation biofuels are generally indistinguishable from their first-generation counterparts, with cellulosic ethanol, for example, being a direct substitute for ethanol made from corn or sugarcane.

The track record with commercialisation of cellulosic biofuels is yet to match the ambitious targets set in the wake of the food and fuel price peaks of 2008. For example, a lack of commercial production in the United States forced the Environmental Protection Agency (EPA) to revise its cellulosic biofuel requirement for 2014 downwards to 17 million gallons, less than 1 per cent of the 1.27 billion gallons that had previously been set as the 2014 target back in 2010 (Environmental

Box 1.2 **Corn ethanol and food versus fuel**

The United States overtook Brazil to become the world's largest ethanol producer in 2006 (Figure 1.5), relying on corn (maize) as its principal feedstock rather the sugarcane used in Brazil. By 2009/10, ethanol production was consuming about a third of the total US corn crop (USDA, 2010).

The sheer size and rapid growth of the US corn ethanol sector has made it a major focus of the so-called *food versus fuel* debate. This debate reached a peak when global food prices reached a record high in 2008, with numerous researchers, policy-makers and commentators weighing in. Jean Zeigler, the UN Special Rapporteur on the right to food, described the diversion of food crops into biofuels as a "crime against humanity", while researchers such as Donald Mitchell of the World Bank concluded that it was the most important factor in the increase in global food crops in the years leading up to 2008 (Mitchell, 2008). In contrast, UN Energy (2007, p. 31) stated that the notion of food versus fuel is "overly simplistic and fails to reflect the full complexity of factors that determine food security at any given place and time" and other researchers emphasised a variety of other contributing factors to higher food prices including poor grain harvests, higher farming input costs, regulatory policies, increased demand for food, changes in diet, exchange rates and speculation (e.g. O'Connell et al., 2009).

The implications for biofuels arising from the food versus fuel debate include an increased focus on food security issues in biofuel policy and an increased research and development focus on so-called *second-generation* biofuels that can be produced from non-edible feedstocks. There has also been a focus on assessing land availability, producing fuel from land unsuitable for food and on integrated food and energy systems based on co-production of food and fuel rather than trading one off against the other (e.g. Elbehri et al., 2013). These issues are discussed further in Chapter 5.

Protection Agency, 2013). First-generation ethanol from corn continued to make up the bulk of the US EPA's 15 billion gallon renewable fuel requirement for 2014.

Some authors go beyond the term *second-generation* and describe other new biofuels from sources such as algae as *third-generation* or even *fourth-generation* (e.g. Singh et al., 2011; Lu et al., 2011). However, the approach taken in this book is to use the term *advanced biofuels* as a catch-all term for cellulosic biofuels, algal biofuels and other biofuels that are not yet in widespread use, rather than trying to work out where these often indistinct generational boundaries lie. Furthermore, it is important to remember that terms such as *second-generation* or *advanced biofuels* should only be applied to cellulosic feedstocks when they are used for liquid biofuels, as these woody and fibrous parts of plants have long been commercially utilised for electricity and heat (e.g. sugarcane bagasse, risk husks, woodchips).

From a sustainability point of view, biofuels produced from cellulosic biomass generally offer greater potential to reduce greenhouse gas emissions and do not have a direct impact on food markets, but it is important to note that they do not eliminate land competition issues entirely. Should cellulosic biofuels become more widespread in coming years, there may be increased pressure to establish new tree or grass crops, which further highlights the need to develop energy cropping systems that enhance rather than degrade the land they are grown on.

Another potential issue that is likely to emerge as advanced biofuels become more widespread is increased competition for feedstock supply between the liquid biofuel and electricity/heat sectors. Increasing interchangeability of feedstocks between liquid and solid biofuels is already apparent. For example, more than 20 per cent of the palm oil imported into the European Union for energy purposes in 2012 was used for electricity and heat rather than biodiesel (Gerasimchuk and Koh, 2013). In the US, the production capacity for cellulosic ethanol that can be produced from waste biomass and crops such as switchgrass rather than corn is predicted to grow by up to 4300 per cent between 2013 and 2016 (Solecki et al., 2013).

As some of the traditional distinctions in the bioenergy sector break down, we may also see sustainability standards that have been designed to deal with liquid biofuels from sugarcane, palm oil and corn (e.g. EU and RSB) being applied to bioenergy produced from woody and fibrous biomass. In addition, we may also see some of the ideas about sustainability that have emerged in relation to woody crops shifting into the sustainability standards that are applied to liquid biofuels. In particular, the idea that woody crops such as willows or mallee eucalypts could be used to actively protect and restore landscapes may work its way into the biofuel standards produced by the EU or RSB. Indeed, despite having clear ramifications for sustainability, the idea that bioenergy crops could help to improve ecosystem health is a notable absence from the current EU and RSB standards, which have been developed in response to concerns around crops such as oil palm having the opposite effect.

Now is indeed an interesting time to be exploring the role that energy crops could play in helping us move towards a more sustainable future. The idea that impacts on land and people can be predicted by the form that bioenergy takes (i.e. biofuels or electricity/heat) will increasingly need to be replaced by a focus on the complex interactions between technologies, markets for energy and food, land use regulations, community attitudes, management strategies and policy settings. Hopefully, this book can offer some insights into how we can approach this sustainable future for energy crops.

Structure of the book

Having introduced the concept of bioenergy sustainability in this chapter and discussed how conceptualisations of it can vary, the next six chapters of the book explore in detail the various sustainability issues that have arisen in relation to bioenergy, and around energy cropping in particular. These issues are dealt with in two parts:

- **Part 2** (Chapters 2–4) covers the relationships between *energy cropping and ecosystem health*, including climate change (Chapter 2), deforestation and land degradation (Chapter 3) and the potential for energy crops to assist with ecological restoration (Chapter 4).
- **Part 3** (Chapters 5–7) covers the *socio-economic dimensions of energy cropping*, with a focus on food security (Chapter 5), issues relating to land rights and the impacts of energy cropping on local communities (Chapter 6) and the economics of energy cropping (Chapter 7).

Scattered throughout these chapters are illustrative examples highlighting environmental, social and economic impacts from across the globe. These include situations where bioenergy production has been linked to environmental destruction or social conflict, such as the links between energy cropping and tropical deforestation in southeast Asia, the impact of US corn ethanol production on global food prices and concerns about large-scale land purchases for biofuels in Africa. Other examples highlight the potential for bioenergy to drive positive change, including the mitigation of dryland salinity in Western Australia, the fight against desertification in the African Sahel region and the potential for energy cropping to increase the economic resilience of farmers trying to adapt to climate change.

Part 4 (Chapters 8–10) brings the book to a conclusion with a focus on moving forward in practical ways that enhance the role that bioenergy cropping can play in creating a sustainable future. This includes a review of policy measures from around the world that have been successful in helping to promote positive forms of bioenergy while restricting those with unacceptable impacts. Chapter 8 brings together innovative policy ideas from countries such as Germany, the UK and the USA before Chapter 9 lays out a potential pathway forward for two case study nations – Australia and Brazil.

References

Anseeuw, W., Wily, L. A., Cotula, L. and Taylor, M. (2012) *Land Rights and the Rush for Land: Findings of the Global Commercial Pressures on Land Research Project*, International Land Coalition, Rome.

Bauen, A., Berndes, G., Junginger, M., Londo, M., Vuille, F., Ball, R., Bole, T., Chudziak, C., Faaij, A. and Mozaffarian, H. (2009) *Bioenergy: A Sustainable and Reliable Energy Source*, IEA Bioenergy Secretariat, Rotorua.

Bennett, D., Simons, J. and Speed, R. (2011) *Hydrological Impacts of Integrated Oil Mallee Farming Systems* Department of Agriculture and Food, Perth.

Biofuels Interagency Working Group (2010) *Growing America's Fuel: An Innovation Approach to Achieving the President's Biofuels Target*, Washington, DC.

Brown, L. R. (2008) *Plan B 3.0: Mobilizing to Save Civilization*, W. W. Norton and Company, New York.

Crutzen, P. J., Mosier, A. R., Smith, K. A. and Winiwarter, W. (2007) "N$_2$O release from agro-biofuel production negates global warming reduction by replacing fossil fuels", *Atmospheric Chemistry and Physics Discussions*, 7: 11,191–205.

Diesendorf, M. (2000) "Sustainability and sustainable development", in Dunphy, D., Benveniste, J., Griffiths, A. and Sutton, P. (eds), *Sustainability: the Corporate Challenge of the 21st Century*, Allen and Unwin, Sydney, 19–37.

Dimitriou, I., Baum, C., Baum, S., Busch, G., Schulz, U., Köhn, J., Lamersdorf, N., Leinweber, P., Aronsson, P., Weih, M., Berndes, G. and Bolte, A. (2011) *Quantifying Environmental Effects of Short Rotation Coppice (SRC) on Biodiversity, Soil and Water*, IEA Bioenergy Task 43, 34p.

Doornbosch, R. and Steenblik, R. (2007) *Biofuels: Is the Cure Worse than the Disease?* Organisation for Economic Co-operation and Development, Paris.

Dovers, S. (2005) *Environment and Sustainability Policy*, Federation Press, Sydney.

Edwards, D. P. and Laurance, W. F. (2011) "Loophole in forest plan for Indonesia", *Nature*, 477, 33.

Eide, A. (2008) *The Right to Food and the Impact of Biofuels (Agrofuels): Advanced Copy*, Food and Agriculture Organization of the United Nations, Rome.

Elbehri, A., Segerstedt, A. and Liu, P. (2013) *Biofuels and the Sustainability Challenge: A Global Assessment of Sustainability Issues, Trends and Policies for Biofuels and Related Feedstocks*, Food and Agriculture Organization of the United Nations, Rome.

Energy Information Administration (2014) *International Energy Outlook 2014 with Projections to 2040*, United States Energy Information Administration, Washington, DC.

Energy Information Administration (2015) "International energy statistics: biofuels production", www.eia.gov/cfapps/ipdbproject/IEDIndex3.cfm?tid=79andpid=79andaid=1 (accessed 9 March 2015).

Environmental Protection Agency (2013) *EPA Proposes 2014 Renewable Fuel Standards, 2015 Biomass-Based Diesel Volume*, United States Environmental Protection Agency, Washington, DC.

Ernsting, A. (2007) "Agrofuels in Asia: fuelling poverty, conflict, deforestation and climate change", *Seedling*, July: 25–33.

European Commission (2010) *Report from the Commission to the Council and the European Parliament on Sustainability Requirements for the Use of Solid and Gaseous Biomass Sources in Electricity, Heating and Cooling*, European Commission, Brussels.

European Commission (2015) "Renewable energy: biofuels – sustainability schemes", http://ec.europa.eu/energy/renewables/biofuels/sustainability_schemes_en.htm (accessed 15 January 2015)

European Parliament and Council of the European Union (2009) "Directive 2009/28/EC", *Official Journal of the European Union*, L140: 16–62.

Future Fuels Forum (2008) *Fuel for Thought: The Future of Transport Fuels: Challenges and Opportunities*, CSIRO, Canberra.

Gerasimchuk, I. and Koh, P. Y. (2013) *The EU Biofuel Policy and Palm Oil: Cutting Subsidies or Cutting Rainforest?* International Institute for Sustainable Development, Winnipeg, OH.

Gibbs, H. K., Johnston, M., Foley, J. A., Holloway, T., ChadMonfreda, Ramankutty, N. and Zaks, D. (2008) "Carbon payback times for crop-based biofuel expansion in the tropics: the effects of changing yield and technology", *Environmental Research Letters*, 3: 1–10.

Greenpeace (2007) *How the Palm Oil Industry is Cooking the Climate*, Greenpeace International, Amsterdam.

Hansen, M. C., Potapov, P. V., Moore, R., Hancher, M., Turubanova, S. A., Tyukavina, A., Thau, D., Stehman, S. V., Goetz, S. J., Loveland, T. R., Kommareddy, A., Egorov, A., Chini, L., Justice, C. O. and Townshend, J. R. G. (2013) "High-resolution global maps of 21st-century forest cover change", *Science* 342: 850–53.

Hartman, J. C., Nippert, J. B., Orozco, R. A. and Springer, C. J. (2011) "Potential ecological impacts of switchgrass (*Panicum virgatum* L.) biofuel cultivation in the Central Great Plains, USA", *Biomass and Bioenergy*, 35: 3415–21.

Hirsch, R. L., Bezdek, R. and Wendling, R. (2005) *Peaking of World Oil Production: Impacts, Mitigation, and Risk Management*, US Department of Energy, Washington, DC.

Holthuijzen, W. A. and Maximillian, J. R. (2011) "Dry, hot and brutal: climate change and desertfication in the Sahel of Mali", *Journal of Sustainable Development in Africa*, 13: 245–68.

IEA Bioenergy (2007) *Potential Contribution of Bioenergy to the World's Future Energy Demand*, IEA Bioenergy Secretariat, Rotorua, New Zealand, 12p.

International Energy Agency (2007) *IEA Energy Technology Essentials: Biomass for Power Generation and CHP*, OECD/IEA, Paris.

IPCC (2014) *Climate Change 2014 Synthesis Report*, Fifth Assessment Report, Intergovernmental Panel on Climate Change, Geneva.

Karp, A. and Halford, N. G. (2011) "Introduction", in Halford, N. G. and Karp, A. (eds), *Energy Crops*, Royal Society of Chemistry, Cambridge, 1–12.

Keiner, M. (2006) "Rethinking Sustainability – Editor's Introduction", in Keiner, M. (ed.), *The Future of Sustainability*, Springer, Dordrecht, 1–15.

Laureysens, I., De Temmerman, L., Hastir, T., Van Gysel, M. and Ceulemans, R. (2005) "Clonal variation in heavy metal accumulation and biomass production in a poplar coppice culture: II. vertical distribution and phytoextraction potential", *Environmental Pollution*, 133, 541-551.

Lawn, P. A. (2001) *Toward Sustainable Development: An Ecological Economics Approach*, Lewis Publishers, Boca Raton, FL.

Lu, J., Sheahan, C. and Fu, P. (2011) "Metabolic engineering of algae for fourth generation biofuels production", *Energy and Environmental Science*, 4, 2451–66.

Marcuse, P. (2006) "Sustainability is Not Enough", in Keiner, M. (ed.), *The Future of Sustainability*, Springer, Dordrecht, 55–68.

McDonough, W. and Braungart, M. (2002) *Cradle to Cradle: Remaking the Way We Make Things*, North Point Press, New York.

Meadows, D. H., Meadows, D. L., Randers, J. and Behrens III, W. W. (1972) *The Limits to Growth: A Report for the Club of Rome's Project on the Predicament of Mankind*, Universe Books, New York.

Mitchell, D. (2008) *A Note on Rising Food Prices*, World Bank Development Prospects Group, Washington, DC.

O'Connell, D., Braid, A., Raison, J., Handberg, K., Cowie, A., Rodriguez, L. and George, B. (2009) *Sustainable Production of Bioenergy: A Review of Global Bioenergy Sustainability Frameworks and Assessment Systems*, RIRDC, Canberra.

Oxfam International (2007) *Bio-fuelling Poverty: Why the EU Renewable-Fuel Target May Be Disastrous for Poor People*, Oxfam, Oxford.

Riddell-Black, D., Toky, O. P., Harris, P. J. C., Srivastava, R. K., Pandey, A. and Vasudevan, P. (2012) "Opportunities to enhance wood fuel yields in semi-arid regions of India using wastewater", 24th Session of the International Poplar Commission, Dehradun, India.

Romijn, H. A. (2011) "Land clearing and greenhouse gas emissions from Jatropha biofuels on African Miombo Woodlands", *Energy Policy*, 39: 5751–61.

Roundtable on Sustainable Biomaterials (2010) *Global Principles and Criteria for Sustainable Biofuels Production: Version 2.1*, Ecole Polytechnique Federale de Lausanne, Lausanne.

Rubin, O. D., Carriquiry, M. and Hayes, D. J. (2008) *Implied Objectives of US Biofuel Subsidies*, Center for Agricultural and Rural Development, Iowa State University, Ames, IA.

Rutovitz, J. and Passey, R. (2004) *NSW Bioenergy Handbook*, NSW Government, Sydney.

Schoneveld, G. C. (2010) *Potential Land Use Competition from First-Generation Biofuel Expansion in Developing Countries*, Center for International Forestry Research, Bogor.

Schroeder, W. (2012) "Capacity of poplar and willow clones to withstand high levels of wastewater application", 24th Session of the International Poplar Commission, Dehradun, India.

Sheil, D., Casson, A., Meijaard, E., Noordwijk, M. v., Gaskell, J., Sunderland-Groves, J., Wertz, K. and Kanninen, M. (2009) *The Impacts and Opportunities of Oil Palm in Southeast Asia: What Do We Know and What Do We Need to Know?* Center for International Forestry Research, Bogor.

Simpson, J. A., Picchi, G., Gordon, A. M., Thevathasan, N. V., Stanturf, J. and Nicholas, I. (2009) *Short Rotation Crops for Bioenergy Systems: Environmental Benefits Associated with Short-Rotation Woody Crops*, Bioenergy Task 30, IEA, Paris.

Singh, A., Olsen, S. I. and Nigam, P. S. (2011) "A viable technology to generate third-generation biofuel", *Journal of Chemical Technology and Biotechnology*, 86: 1349–53.

Solecki, M., Scodel, A. and Epstein, B. (2013) *Advanced Biofuel Market Report 2013: Capacity through 2016*, Environmental Entrepreneurs, San Francisco, CA.

Stucley, C., Schuck, S., Sims, R., Bland, J., Marino, B., Borowitzka, M., Abadi, A., Bartle, J., Giles, R. and Thomas, Q. (2012) *Bioenergy in Australia: Status and Opportunities*, Bioenergy Australia, Killara.

Swamy, S. L. (2012) "Mitigation and adaptation strategy to climate change – a case study of *Populus deltoides* based agroforestry system in Chhattisgarh, Central India", 24th Session of the International Poplar Commission, Dehradun, India.

UN Energy (2007) *Sustainable Bioenergy: A Framework for Decision Makers*, United Nations, New York.

United Nations (2015) *The Millennium Development Goals Report 2015*, United Nations, New York.

United Nations Conference on Environment and Development (1992) *Rio Declaration on Environment and Development*, United Nations Conference on Environment and Development, Rio de Janeiro.

USDA (2010) *USDA Agricultural Projections to 2019*, United States Department of Agriculture, Washington, DC.

Weih, M. (2012) "Poplar and willow biomass from marginal land? Production, ecological and environmental implications", 24th Session of the International Poplar Commission, Dehradun, India.

World Commission on Environment and Development (1987) *Our Common Future*, Oxford, Oxford University Press.

Part 11

Energy cropping and ecosystem health

Chapter 2

Bioenergy and climate change

Climate change is perhaps the sustainability issue that dominates discussions of bioenergy more than any other. Along with its social and economic dimensions, climate change has serious implications for the health of Earth's ecosystems. In its Fifth Assessment Report released in 2014, the Intergovernmental Panel on Climate Change (IPCC) highlighted a range of risks to ecosystem health, including increased rates of species extinction, flooding of low-lying habitats and threats from temperature rises and ocean acidification in sensitive areas such as coral reefs and polar ecosystems (IPCC, 2014). Mitigating climate change by replacing fossil fuels with low-carbon fuels has been a major motivation behind the promotion of bioenergy by governments around the world, including in the EU, US and Australia. However, other authors have raised concerns that certain forms of bioenergy, particularly liquid biofuels, have the potential to exacerbate climate change risks (e.g. Ernsting et al., 2007; Crutzen et al., 2007; Searchinger et al., 2008).

Attempting to understand the diverging views around bioenergy and climate change can be a daunting task, requiring one to enter a world of complex terms, concepts and acronyms, including:

- *life cycle assessment* (LCA),
- *greenhouse gas savings,*
- *carbon debt,*
- *carbon payback,*
- *indirect land use change* (iLUC) and
- *carbon-negative biofuels.*

The aim of this chapter is to demystify some of these terms while exploring the circumstances under which the competing claims around bioenergy and climate change may hold true. However, as with many aspects of bioenergy sustainability, it is difficult to produce a single uncontested vision of the role that bioenergy can play with regard to the mitigation or exacerbation of climate change. Perspectives on the relationship between bioenergy and climate change will inevitably vary depending on the context in which bioenergy is being used, the assumptions and methodologies used to assess impacts, the level of uncertainty surrounding impacts and the values held by the diverse range of stakeholders with an interest in this topic.

Determining the climate change impacts of bioenergy: the role of life cycle assessment

Viewed narrowly, bioenergy may be seen as *carbon neutral* in the sense that the carbon dioxide (CO_2) emitted during combustion of the fuel is equal to that taken out of the atmosphere during the growth phase of the feedstock. If these two movements of carbon (i.e. from the atmosphere to the plants and from the biofuel back to the atmosphere) are all that is taken into account, one may come to the conclusion that a biofuel like ethanol or biodiesel has no net greenhouse gas (GHG) emissions. Using such a fuel to replace a fossil fuel like gasoline or diesel could therefore be assumed to result in a 100 per cent reduction in baseline GHG emissions (i.e. a 100% GHG *saving*). However, these basic movements of carbon are not the only GHG changes that are relevant to the production of a biofuel. Across the full life cycle of the fuel, there are also likely to be a range of *process emissions* (i.e. GHG emissions from planting, fertilising, harvesting, processing and transporting feedstocks), as well as non-CO_2 GHG gases that may be emitted during combustion (e.g. methane, nitrous oxide). In addition, there may be emissions related to land use change, such as clearing forested land in order to grow the energy crop. Once these sources of emissions are taken into account, the GHG savings from replacing fossil fuels with bioenergy are often much less than 100 per cent.

Accounting for the various sources of emissions throughout the life cycle of a biofuel requires the use of life cycle assessment (LCA), which is "a systematic evaluation of environmental impacts arising from the provision of a product or service" (Horne et al., 2009, p. 2). Many LCAs follow the standardised process laid down in ISO 14040 (International Organization for Standardization, 2006), which divides LCA into four phases:

- goal and scope definition;
- inventory analysis;
- impact assessment; and
- interpretation.

These phases are shown in Figure 2.1, along with examples of the key actions required for a bioenergy LCA. The phases are not strictly sequential, as earlier stages may need to be revisited as results emerge (e.g. the interpretation phase may identify a significant source of emissions that has been left out of the original scope).

The first phase of a LCA involves articulating factors such as why it is being undertaken, what units will be used to weigh up the key impacts of concern, and how far the boundaries of the LCA will extend. In the case of bioenergy and climate change, the goal is typically to determine the net GHG emissions associated with the biofuel's production and use. However, other goals can also come into play, such as determining whether a biofuel is eligible for a given support programme, such as the EU's Renewable Energy Directive (RED). The functional unit can vary but is commonly expressed as a unit of energy (e.g. megajoules or kilowatt-hours) rather than volume or weight, as different fuels will contain different amounts of energy per litre or kilogram. The unit of impact is usually expressed as grams (or kilograms) of carbon dioxide-equivalent emitted for every unit of energy contained in the fuel (e.g. gCO_2-e/MJ).

The use of CO_2-e allows different types of GHG emissions to be compared side-by-side (e.g. one kg of methane = 21 kg CO_2-e). Results may also be expressed as the percentage of GHG saving compared to using a fossil fuel, which requires not only data on the life cycle emissions of the biofuel, but also on those of the fossil fuel to which it is being compared.

Another key step in the goal and scope definition phase is the setting of boundaries around which emission sources will be included in the LCA. This can be one of the most controversial elements of bioenergy LCAs, with some aspects such as cultivation, transport and processing almost always included, but other sources of emissions that are less directly linked to the biofuel often left out. Land use change is often excluded from biofuel LCAs, particularly indirect land use change (discussed later in the chapter).

The *inventory analysis* phase for a bioenergy LCA involves mapping out all the inputs and outputs that occur within the boundary that has been set. For example, a key input for many energy cropping systems is nitrogen-based fertiliser, which

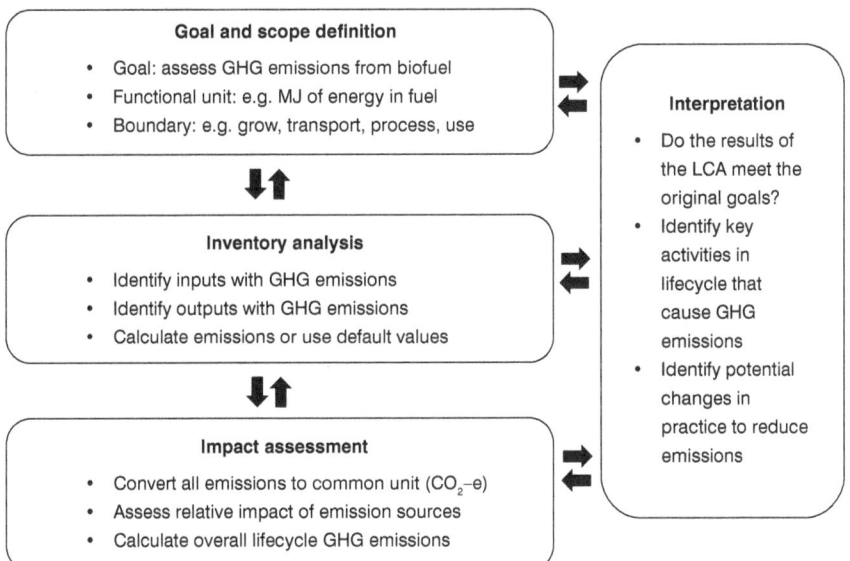

Figure 2.1 Phases of LCA with examples of actions required for a bioenergy LCA

can in turn produce nitrous oxide (a potent greenhouse gas) as an output that is emitted from the cropland to the atmosphere (Crutzen et al., 2007). The *impact assessment* phase of LCA involves converting these various inputs and outputs into carbon dioxide equivalents in order to determine the relative impact of each life cycle stage on the overall emissions profile of the biofuel.

The final phase for a bioenergy LCA is *interpretation*, which involves a strategic assessment of the LCA results, including consideration of any changes that could be made to the production process in order to reduce overall emissions. An example of this can be seen in a recent study by the Malaysian Palm Oil Board undertaken in response to EU and US concerns about the life cycle GHG emissions of palm oil biodiesel (Subramaniam et al., 2014). In this study, interpretation of LCA results across a range of scenarios identified methane emitted from palm oil mill effluent as the single most significant source of emissions for palm oil production (not including land use change). The study concluded that a switch to 100 per cent methane capture at Malaysian palm oil mills could reduce life cycle GHG emissions by more than 50 per cent.

A key challenge for biofuel producers seeking to undertake GHG LCAs in order to demonstrate compliance with GHG saving requirements is obtaining accurate and verifiable emissions data. To assist with this task, much work has been undertaken by various government and non-government institutions to identify default values and develop calculators to assess the GHG emissions associated with the various life cycle stages of common biofuels. Box 2.1 outlines how these default values can be used to test the eligibility of biofuels under the EU's Renewable Energy Directive (RED).

As in the EU, the US Environmental Protection Agency (EPA) has assisted biofuel producers by undertaking its own LCAs, making it easier to demonstrate compliance with the Renewable Fuel Standard (RFS). The RFS promotes biofuels by mandating their use by fuel refiners and importers, with different GHG saving requirements set for three different categories of renewable fuel. Renewable fuel produced at new facilities (post-2007) must achieve a 20 per cent saving, biomass-based diesel and advanced biofuel require a 50 per cent saving and cellulosic biofuel requires a 60 per cent saving. In 2010, the EPA published a rule stating that a range of fuels and fuel pathways comply with the relevant thresholds. Ethanol produced from corn starch at new efficient facilities (if natural gas, biomass or biogas is used for process energy) is considered compliant with the 20 per cent threshold, biodiesel from soy oil, waste oils or algal oil is considered compliant with the 50 per cent threshold, and a range of cellulosic feedstocks and pathways, including both waste biomass and purpose-grown energy crops are considered compliant with the 60 per cent threshold (Environmental Protection Agency, 2010). Unlike the EU, the US EPA's LCA methodology includes emissions from indirect land use change (iLUC).

Neither the EU nor the US extend their GHG saving requirements to solid or gaseous forms of bioenergy used for electricity or heating. The justification given for this differential treatment in the EU is that solid biomass fuels used for

Box 2.1 **Assessment of life cycle greenhouse gas emissions under the EU's Renewable Energy Directive**

The EU Renewable Energy Directive (RED) is an example of a policy tool that incentivises the production of biofuels while restricting its incentives to biofuels that meet basic sustainability standards. The RED aims to ensure that 20 per cent of the EU's overall energy supply comes from renewable sources by 2020, including 10 per cent of the energy used for road transport. Within these broader targets, each EU member state has its own target, which may be achieved through the use of various incentives, such as the Renewable Transport Fuel Obligation in the UK. As discussed in Chapter 1, biofuels can only be counted towards the road transport fuel target if they achieve at least 35 per cent GHG savings across their lifecycle compared with fossil fuels and are not produced from land with high biodiversity values, high carbon stocks or underlying stores of peat. The 35 per cent threshold is scheduled to rise to 50 per cent in 2017 and 60 per cent in 2018.

To help biofuel suppliers calculate GHG savings, the RED specifies a methodology and lists *default* values for the emissions associated with the various lifecycle stages of common biofuels. The equation specified for the calculation of total lifecycle emissions is as follows:

$$E = e_{ec} + e_l + e_p + e_{td} + e_u - e_{sca} - e_{ccs} - e_{ccr} - e_{ee}$$

Where:

E = total emissions from the use of the fuel

e_{ec} = emissions from the extraction or cultivation of raw materials

e_l = annualised emissions from carbon stock changes caused by land-use change

e_p = emissions from processing

e_{td} = emissions from transport and distribution

e_u = emissions from the fuel in use

e_{sca} = emission saving from soil carbon accumulation via improved agricultural management

e_{ccs} = emission saving from carbon capture and geological storage

e_{ccr} = emission saving from carbon capture and replacement

e_{ee} = emission saving from excess electricity from cogeneration

The above equation, like all LCA methodologies, involves boundary-setting. In this case, emissions are not included from certain activities that are less directly linked to the production of the biofuel, such as the manufacture of equipment and, most controversially, indirect land use

Continued...

Box 2.1 continued

change (iLUC). The methodology also states that emissions from the fuel in use (e_u) can be taken to be zero (i.e. the CO_2 emitted during combustion is equal to that taken up by the plant during growth). This simplifies calculations, but excludes emissions of methane and nitrous oxide that may occur due to incomplete combustion.

Annex V to the RED provides *default* emissions values for common biofuels covering cultivation, processing and transport and distribution. A biofuel supplier may choose to use these default values rather than calculate actual emissions values for each stage of their biofuel's lifecycle. No default values are provided for emissions from land use change, soil carbon accumulation or carbon capture, meaning that biofuel suppliers must calculate these values themselves (although in practice they are often zero). Annex V also provides default values for GHG emissions from fossil fuels that are displaced by liquid biofuels (83.8 g CO_2-e/MJ), which can be used to convert the total emissions from the biofuel into a percentage GHG saving. The fossil fuel values include not only the GHG emissions released through combustion, but also emissions associated with extraction, processing and transport (i.e. a *well-to-wheels* approach).

The RED provides an additional mechanism to streamline the process of calculating percentage GHG savings by allowing voluntary sustainability schemes to be recognised as complying with EU requirements and methodologies. As of January 2014, nineteen schemes had been recognised under this provision, allowing biofuel suppliers to cite certification under these schemes as proof of compliance with the RED's sustainability criteria. Prominent schemes include the Roundtable on Sustainable Biofuels (RSB), Roundtable on Sustainable Palm Oil (RSPO) and BioGrace (which only covers GHG savings and not the other sustainability criteria).

electricity and heating tend to have higher GHG savings because they are often sourced from waste biomass and employ processing techniques with low energy consumption, such as pelletisation (European Commission, 2010).

Solid biomass also has an additional incentive in the EU compared to liquid biofuels, stemming from the EU Emissions Trading Scheme. This cap-and-trade scheme requires electricity generators to hold permits covering their GHG emissions (priced at around €5–10 in early 2015), which provides an advantage for biomass-fired plants over those using fossil fuels such as coal. While liquid biofuels that are used for electricity generation also benefit from this scheme, the majority of liquid biofuel use is for road transport, which is not covered by the ETS. All solid biomass is "zero-rated" under the ETS (i.e. assumed to have zero GHG emissions), while liquid biofuels used for electricity generation are also "zero-rated" if they meet the RED sustainability criteria, including in relation to life-cycle GHG savings (European Commission, 2012a).

The argument that solid biomass fuels have greater GHG savings is broadly supported by the data shown in Figure 2.2, which compares GHG savings for common solid biomass fuels used for electricity in the EU with the GHG savings for selected liquid biofuels listed in Annex V of the RED.

In general, the solid biomass fuels shown in Figure 2.2 have a higher GHG saving than the liquid biofuels shown. In particular, solid fuels that take the form of woodchips or are produced from forestry and agricultural wastes tend to have the highest GHG savings (60–90%). Wood pellets, which require further processing, have lower GHG savings than woodchips. Eucalypts grown using short-rotation coppicing (SRC) show the lowest GHG savings among the solid fuels in Figure 2.2, and in their pelleted form are the only solid fuel shown that would fail to meet a 35 per cent benchmark (if one were applied to bioelectricity). The main reasons for the poor GHG saving for SRC eucalypts relative to the other solid fuels are longer transport distances (import is assumed from tropical countries) and higher N_2O emissions from nitrogen fertiliser (JRC, 2014). In comparison, SRC poplar, which is assumed to be unfertilised and sourced locally, delivers very similar GHG savings to cultivated stemwood, forest residues and agricultural residues. The use of solid biomass for heating, which is not shown in Figure 2.2, tends to result in slightly higher GHG savings than the results shown for bioelectricity (JRC, 2014).

In comparison to the solid fuels, the liquid biofuels shown in Figure 2.2 tend to have lower GHG savings. Three of the nine biofuels shown (not including advanced biofuels) would fail to meet the EU's 35 per cent benchmark, while all but two (sugarcane ethanol and waste oil biodiesel) would fail to meet the higher 60 per cent benchmark that is due to be applied from 2018. In contrast, all four of the advanced (second-generation) biofuels, which are not yet in widespread use, would meet the 60 per cent threshold.

When considering the results shown in Figure 2.2, it should be remembered that these GHG savings are based on particular methodologies and assumptions favoured by the EU. Different analyses may yield different results. The default values published by the EU are deliberately set at conservative levels to ensure that producers of poor performing biofuels cannot gain an advantage by choosing to use the default value over their own calculated value. The EU has also published a list of *typical* GHG savings, which are higher than the default values in many cases.

The conservative nature of the EU defaults can be seen by comparing them to the GHG savings estimated by Langeveld et al. (2014) for EU ethanol from wheat (23–63% as opposed to 16%) and rapeseed biodiesel (40–44% as opposed to 38%). In this case, a difference in methodology plays a key role, with Langeveld et al. (2014) applying a fossil fuel reference that is higher than the EU's (90 rather than 83.8 g CO_2-e/MJ). A recent report by van den Bos and Hamelinck (2014) argues for an even higher fossil fuel reference of 115 g CO_2-e/MJ on the basis that biofuels serve to prevent the extraction of new sources of unconventional oil, such as shale oil, which require much higher inputs of energy than traditional oil extraction.

Notwithstanding the debate around fossil fuel reference values, the data shown in Figure 2.2 provides strong evidence that the climate change mitigation potential of biofuels would be enhanced by a shift away from first-generation biofuels produced from common agricultural crops towards biofuels produced from wastes or cellulosic feedstocks. Moreover, there is also a strong argument for supporting the use of solid biomass fuels for electricity and heat, as this generally results in strong GHG savings and the feedstocks involved

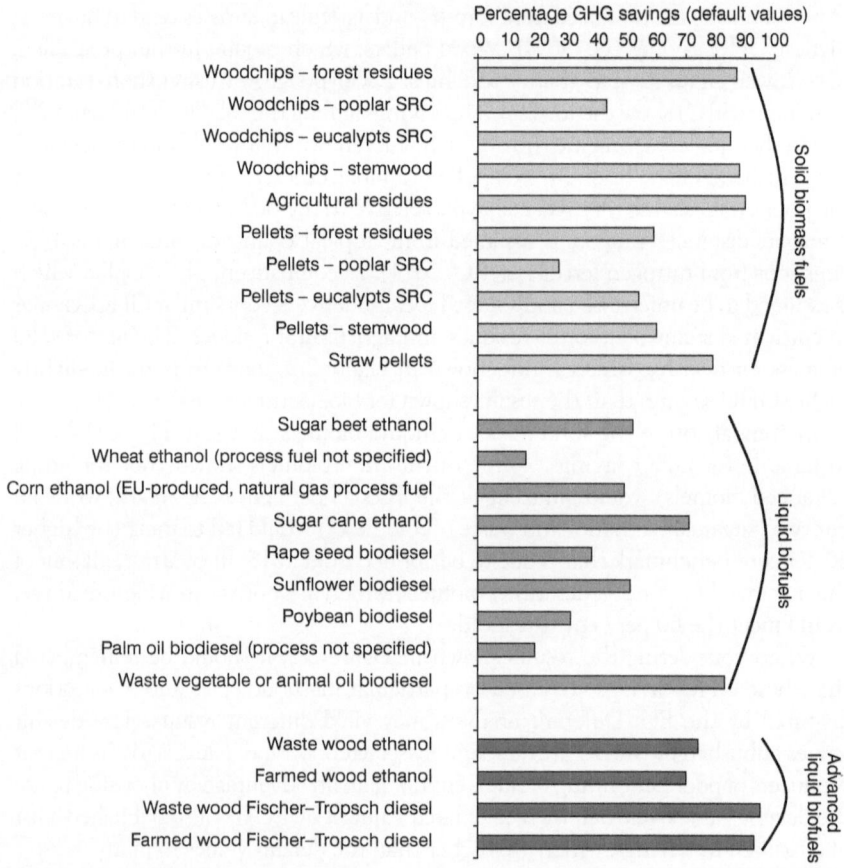

Figure 2.2 Comparison of default GHG savings for selected solid biomass fuels used for electricity and selected liquid biofuels used for transport in the EU

Source: Solid biomass values from JRC (2014) are based on the same methodology used to determine default GHG savings for liquid biofuels under Annex V of the RED (European Parliament and Council of the European Union, 2009)

Notes: All values assume zero emissions from land use change. SRC = short rotation coppice. All solid biomass values assume a transport distance to the point of combustion of 0–500 km except for eucalyptus SRC, which is assumed to have a transport distance of 2,500–10,000 km due to import from tropical areas. Poplar SRC is assumed to be unfertilised. All woodchip values assume wood is used as a process fuel.

are largely the same as those used for cellulosic biofuels (i.e. agricultural and forestry residues and woody or fibrous energy crops). However, there are some important exceptions, both in terms of first-generation biofuels that are capable of offering strong GHG savings (e.g. sugarcane ethanol) and solid fuels that have a poor GHG performance (e.g. imported eucalyptus pellets). Furthermore, there is another key factor that has so far been left out, but which can have significant impacts on GHG savings – land use change.

Factoring in land use change

All the GHG savings figures quoted so far in this chapter assume that no greenhouse gases are emitted from changes in land use for the purpose of biofuel feedstock production. In reality, a change in land use can reduce the amount of carbon stored in the land unit through clearing trees, burning vegetation and disturbing soil carbon, with the lost carbon making its way back into the atmosphere as CO_2 (or potentially methane). The magnitude of this change depends on the amount of carbon stored in the soils and biota before conversion to energy cropping compared to the amount stored after conversion. Stored carbon is likely to decrease most substantially when forests are cleared and peatlands are drained, releasing methane. However, it is also important to remember that stored carbon levels can actually increase when woody perennial crops are planted on degraded land.

A key difference between land use change emissions and process emissions (i.e. those from cultivation, transport and processing) is that land use change generally represents a one-off change in carbon stocks that occurs prior to bioenergy production. Unlike process emissions, which are ongoing and can be allocated to each unit of biofuel with relative ease, allocating land use change emissions to each unit of biofuel requires the emissions to be amortised, or spread over a period of years in which biofuel production is likely to occur. The standard figure adopted by the EU, RSB and others is 20 years. Once these one-off emissions have been spread over 20 years' worth of biofuel production, they can then be added to the process emissions and compared to a fossil fuel reference to produce a GHG saving value that includes land use change.

An alternative approach to accounting for land use change involves excluding these emissions from GHG savings and instead determining how many years of biofuel production (with associated GHG savings from replacing fossil fuels) would be required to "pay back" the "carbon debt" incurred through the initial land use change. For example, Gibbs et al. (2008) calculated that, if tropical peat forests are cleared to produce palm oil for biodiesel (releasing both CO_2 from the forest and methane from the peat), the land would need to be used for biodiesel production for 918 years in order to pay back the emissions resulting from land use change. This represents an extreme case, but is often the kind of biofuel production system that stimulates headlines proclaiming that biofuels are "worse than fossil fuels" (e.g. BBC News, 2013; Guardian, 2013). Table 2.1 attempts to give a more representative view by showing how the inclusion of land use change emissions affects the GHG savings of selected first-generation EU biofuels from Figure 2.2.

Before discussing the examples shown in Table 2.1, it is important to note that we are only including *direct* land use change (i.e. relating to land that is directly being used for energy cropping) at this point. *Indirect* land use change will be dealt with shortly. It is also important to note that the emissions values for land use change in Table 2.1 are based on generalised data rather than specific case studies of biofuel production. In practice, the impacts of land use change can be highly variable, even within the same broad land type in the same country. Biofuels produced from wastes (e.g. biodiesel from used cooking oil) and feedstocks grown on existing cropland are excluded from the table, as these options are assumed to result in no land use change. Advanced biofuels are also excluded (e.g. ethanol from farmed wood), as these are not yet common in the EU and are not covered by the UK Government dataset used for these calculations (Renewable Fuels Agency, 2010).

The examples given in Table 2.1 highlight the highly variable nature of land use change emissions and the impacts they can have on overall GHG savings. Four of the eight biofuels that have been chosen show negative GHG savings once land use change is factored in, indicating that they are indeed worse than

Table 2.1 Impact of land use change emissions on percentage of greenhouse gas savings for selected EU biofuels

Biofuel	GHG savings (excl. land use change)	Land assumed to be converted for feedstock[a]	GHG savings including land use change[b]	Payback period (years)
Sugar beet ethanol	52%	Grassland (UK)	13%	15
Corn ethanol (natural gas process fuel)	49%	Grassland (France)	−56%	43
Wheat ethanol (process fuel not specified)	16%	Forest (Germany)	−351%	446
Sugar cane ethanol	71%	Grassland (Brazil)	71%	0
Rapeseed biodiesel	38%	Grassland (Canada)	5%	18
Soybean biodiesel	31%	Grassland (USA)	−12%	28
Palm oil biodiesel (methane not captured)	19%	Forest (Malaysia)	−174%	205
Palm oil biodiesel (methane captured)	56%	Degraded land (Malaysia)	90%	-

Notes:

[a] Land use change emissions for each land category and biofuel from UK Government (Renewable Fuels Agency, 2010), except for the final entry which applies the EU bonus of 29 g CO_2-e/MJ for planting on degraded land (European Parliament and Council of the European Union, 2009).

[b] Land use change emissions have been amortised over a 20-year period.

using fossil fuels. Interestingly, the worst performer is not palm oil from cleared Malaysian forests, but rather wheat ethanol from cleared German forests. This is not a reflection of the amount of carbon stored in each forest type, but has more to do with the relative productivities of each cropping system (i.e. the carbon debt is repaid more quickly in the case of palm oil due to a higher rate of biofuel production). Sugar beet ethanol and rapeseed biodiesel are each able to maintain a small GHG saving when produced on converted grassland, but these savings are well below the 35 per cent EU threshold.

Sugarcane ethanol from Brazilian grassland maintains its high GHG saving of 71 per cent in Table 2.1 even when land use change is included, indicating that the conversion of grassland to sugarcane in this location would result in no net loss of carbon. This is supported by Lange (2011), who estimates a lifecycle GHG saving of 80 per cent for sugarcane ethanol grown on converted Brazilian grassland and suggests that such a change would actually increase the stored carbon in the landscape.

The idea that bioenergy crops could increase stored carbon levels is also demonstrated by the final entry in the table, which shows a kind of "best-case scenario" for palm oil that is quite different to the negative portrayals discussed so far. Firstly, methane from milling wastes is assumed to be captured in this scenario, resulting in a greatly improved GHG saving (56%) before land use change is even considered. Secondly, the oil palm is assumed to be planted on degraded land, which would have very little stored carbon in its vegetation and soils. The use of degraded land qualifies for a "bonus" GHG saving under EU rules and leads to an overall saving of 90 per cent, which is comparable to some of the best-performing advanced biofuels from Figure 2.2.

Figure 2.3 demonstrates how energy cropping systems can increase carbon levels in a land unit over time, using a hypothetical woody crop as an example. Even with periodic harvest, Figure 2.3a shows how the level of stored carbon in the land unit can increase over time before reaching an average level that is higher than that which prevailed prior to planting. A particularly beneficial feature of short rotation coppice (SRC) systems involving crops like willows or eucalypts is that the coppicing process leaves the roots and stools behind and does not require the soil to be tilled after every harvest. Lockwell et al. (2012) report on a real-world example of a willow crop in Quebec, Canada, that showed an increase in soil carbon over time and resulted in higher levels than a comparison site under annual crops. However, while there is a general consensus that SRC and perennial grass crops can sequester more soil carbon than annual crops, Lowrance and Davis (2014) also point out some limitations relating to SRC crops. These include that soil carbon will not increase indefinitely (only until it reaches a new equilibrium), that changes in soil carbon may be difficult to verify as much as a decade after planting, and that the soil carbon benefits of bioenergy crops are not as clear-cut when grasslands are replaced rather than annual crops.

In cases where the increase in stored carbon is a one-off (e.g. converting degraded land to a woody perennial crop), the increase in sequestered carbon

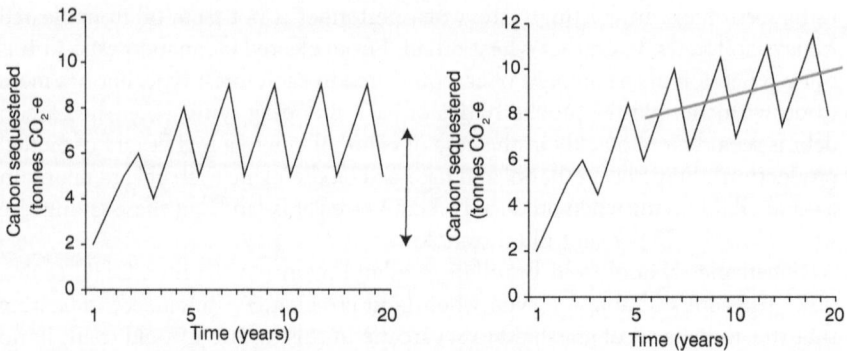

Figure 2.3 Hypothetical examples of SRC crops increasing levels of stored carbon. (*a*) One-off increase in average levels of stored carbon (once a new equilibrium is reached). (*b*) Ongoing increase in stored carbon from a bioenergy system that involves the biochar from the biomass being sequestered in soils (not necessarily on the same land unit that the crop is grown)

should be amortised over 20 years before being factored into life cycle calculations, just as one-off losses of carbon from deforestation must be spread over 20 years. However, Figure 2.3*b* shows how it is also possible to produce bioenergy crops that sequester carbon on an ongoing basis by adding *biochar* to the soil. Biochar is a stable form of solid carbon produced from biomass through pyrolysis, which has shown the potential to reduce GHG emissions from soils and increase soil carbon in a variety of contexts (e.g. Vaccaria et al., 2011; Augustenborg et al., 2012; Kammann et al., 2012). Another option for achieving an ongoing increase in stored carbon is geosequestration of CO_2 gas from large-scale bioenergy facilities, which could be applied in much the same way as for coal-fired power stations.

Bioenergy cropping systems that sequester carbon create the possibility of producing biofuels that go beyond carbon neutral and are actually *carbon-negative* (Mathews, 2008). This can occur when the increase in stored carbon levels is sufficient to outweigh all other GHG emissions from the production of bioenergy (i.e. from cultivation, transport and processing). Lange (2011) cites sugarcane ethanol grown on degraded land in Brazil and palm oil biodiesel grown on degraded land in southeast Asia as two biofuels that are capable of qualifying as carbon-negative, with GHG savings greater than 100 per cent. As advanced biofuels from cellulosic tree and grass crops become more widespread, it is likely that more examples of carbon-negative biofuels will emerge.

Indirect land use change

Indirect land use change (iLUC) is a term used to describe land use changes that occur elsewhere as a consequence of a bioenergy project (Berndes et al., 2011). Unlike the direct impacts discussed so far, iLUC does not take place on the land unit that is actually being used for bioenergy, but rather on land in an entirely different location, driven by changes in supply, demand and market prices for biofuel feedstocks. However, while the mechanism is different, the impacts are much the same as direct land use change, with carbon lost to the atmosphere when forests and other land are cleared for cropping. The scale of iLUC impacts can vary from the local, such as a small shift of an agricultural frontier into primary forest, to the global, such as corn ethanol production contributing to rising world prices for corn and creating an incentive to clear land on the other side of the world.

Concerns around iLUC are closely connected to issues of food vs fuel (discussed in Chapter 5), as well as deforestation and dispossession in developing countries (discussed in Chapters 3 and 6). These concerns rose to prominence as global food and oil prices were heading towards their 2008 peaks. In a prominent and controversial paper, Searchinger et al. (2008) estimated that full life cycle GHG emissions from US corn ethanol are actually 93 per cent higher than emissions from gasoline if iLUC is included. In response to these concerns, the US EPA undertook its own GHG calculations for corn ethanol, finding that, even with iLUC included, life cycle emissions were still 20 per cent lower than the life cycle emissions of gasoline (Environmental Protection Agency, 2010).

The difference between the US EPA result and that of the Searchinger paper is largely due to differing assumptions, such as whether corn yields would increase in response to higher prices, the extent to which ethanol by-products could be added to markets for animal feed and the types of land that would be converted for corn production globally. Berndes et al. (2011, p. 53) cite a range of other studies that have also calculated much lower iLUC emissions from US corn ethanol than those published by Searchinger et al. and conclude that "short-term emissions from land use change are not sufficient reason to exclude bioenergy from the list of worthwhile technologies for climate change mitigation".

While the US Government has decided to include iLUC emissions in its LCA calculations, such an approach has not yet been adopted by the EU or the Roundtable on Sustainable Biomaterials (RSB). However, an EU proposal drafted in 2012 would, if adopted, require iLUC emissions to be included in the life cycle GHG calculations for biofuels (European Commission, 2012b). The estimated iLUC emissions that would need to be included under the proposal are 12 g CO_2-e/MJ for cereals and other starch-rich crops, 13 g CO_2-e/MJ for sugars and 55 g CO_2-e/MJ for oil crops. Figure 2.4 demonstrates how the inclusion of these iLUC emissions would affect overall GHG savings, using same selection of EU biofuels as for Table 2.1. The negative GHG savings for rapeseed, soybean and oil palm when iLUC is included demonstrate how oil crops would be particularly affected by the proposed changes.

Comparing the GHG savings in Figure 2.4 (including iLUC) to the GHG savings in Table 2.1 (including direct land use change) produces mixed results. In the case of rapeseed biodiesel, soybean biodiesel and sugarcane ethanol, the inclusion of iLUC emissions has a bigger impact on overall GHG savings than the inclusion of the direct land use changes assumed for Table 2.1. For ethanol from sugar beet, corn and wheat, the impacts of including iLUC are less than the impacts of the direct land use changes assumed for Table 2.1. For palm oil biodiesel, the result depends on whether the land assumed for direct land use change is forest or degraded land. When performing life cycle GHG calculations, it is not necessary to include both direct and indirect land use change emissions for the same biofuel, as direct emissions occur when forests or grasslands are displaced and indirect emissions occur when other crops are displaced.

One of the key reasons that the 2012 EU proposal relating to iLUC emissions has not yet been adopted is the level of uncertainty associated with available iLUC models. Uncertainty was cited as a key concern in a 2010 report to the European Commission on iLUC (Al-Riffai et al., 2010), as well as by the UK Government (Department for Transport, 2012). Apart from this uncertainty, including iLUC emissions also raises ethical and philosophical questions about when it is appropriate to attribute a given impact to a given crop. The RSB points out that biofuel certification alone may not be effective in combating iLUC due to these impacts being "beyond the control of the individual farmer or biofuels producer" (Roundtable on Sustainable Biomaterials, 2010, p. 4). It is also

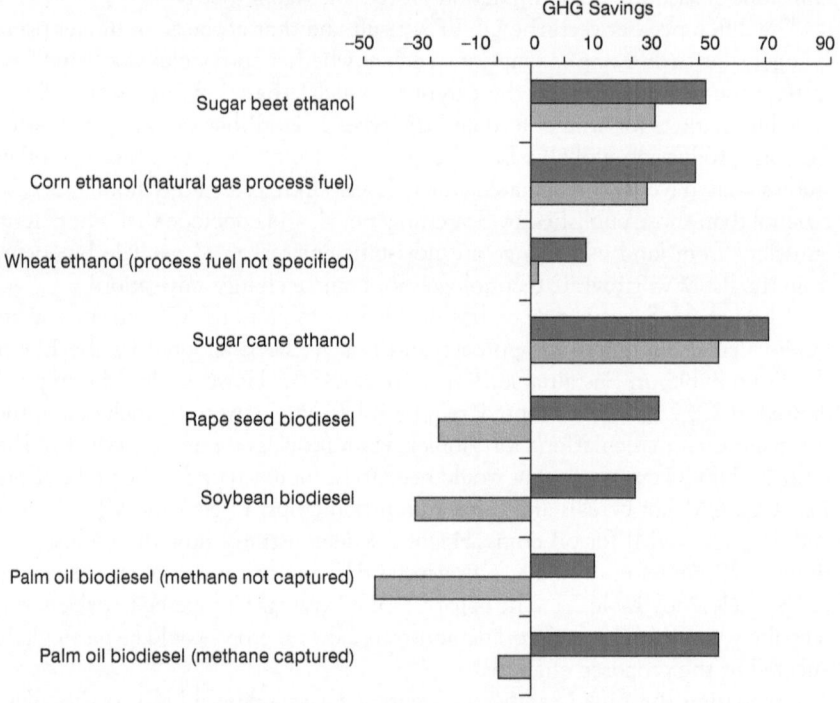

Figure 2.4 Life cycle GHG savings of selected EU biofuels with and without iLUC

Source: Estimated iLUC emissions from European Commission (2012b). Default values excluding land use change from European Parliament and Council of the European Union (2009)

important to remember that demand for food, feed and fibre is rising alongside demand for biofuels, and that agricultural frontiers are expanding as a result of the combined demand for all of these products. Zilberman et al. (2010) point out that including iLUC emissions in biofuel LCAs assigns responsibility for these impacts to the biofuels that have indirectly contributed to them rather than addressing land use change directly in whichever industry sector it occurs (i.e. for food, feed, fibre or biofuels). They argue that we shouldn't settle for such a "second-best" solution. Further discussion of the interconnected nature of food, feed, fibre and bioenergy markets is provided in Chapter 5, including an overview of some of the land availability models that have been developed.

Implications for bioenergy crops

Taken together, the issues, statistics and policy measures covered in this chapter help to forge a vision of the role that bioenergy can play in the mitigation of climate change going into the future. Given that this book is primarily concerned with energy crops, it is important to consider how they fit into this vision. It is clear that many of the common agricultural crops that form the feedstocks for first-generation biofuels, such as corn, wheat and soybean will struggle to satisfy the increased GHG saving expectations that have been set by the EU, the US and others. These feedstocks showed some of the lowest GHG savings in Figure 2.2 and also pose a significant risk from land use change (both direct and indirect). Sugarcane and palm oil may be exceptions due to their high levels of productivity and low fertiliser requirements, but careful attention will be required to ensure that they do not contribute to undesirable land use change, and in the case of palm oil, that any methane from milling wastes is captured and converted to CO_2 (preferably with productive use for electricity or heating).

While many of the energy crops grown for first-generation biofuels have been associated with high GHG emissions, other energy crops have shown the potential to achieve strong GHG savings or even to produce carbon-negative biofuels. As discussed previously, these high-performing forms of bioenergy include advanced biofuels from cellulosic feedstocks, electricity and heat from cellulosic feedstocks, and perennial crops that are able to increase the amount of stored carbon in the landscape. The EU and the US have each introduced measures to preferentially promote forms of bioenergy that offer higher GHG savings. The US approach is to set fuel volume requirements that aim to not only increase the volume of advanced biofuels in use, but also to increase the proportion of renewable fuels that are made up of advanced biofuels (Figure 2.5). In practice, these ambitious targets have not yet been met and advanced biofuel requirements have had to be lowered in line with available fuel volumes (Environmental Protection Agency, 2013).

The EU has sought to promote biofuels with higher GHG savings by progressively increasing its GHG savings requirements from 35 per cent in 2009 to 50 per cent in 2017 and 60 per cent in 2018. The amendments to the RED proposed in 2012 would accelerate this transition, with all new facilities built after 2014 being required to achieve 60 per cent GHG savings. The EU also provides an additional incentive for biofuels produced from wastes or cellulosic feedstocks by allowing them to be "double-counted" against the biofuel targets of member states. When this provision is implemented at the national scale, such as under the UK's Renewable Transport Fuel Obligation (RTFO), a fuel supplier has twice as much incentive to supply these types of biofuels because they can meet their biofuel obligation with half as much biofuel. The proposed changes to the RED would expand on the double-counting principle by allowing biofuels from waste feedstocks to be counted for up to four times their actual energy content and those from cellulosic energy crops to be double-counted. In addition, a cap would be placed on the proportion of the EU's transport fuel target that could be met by first-generation biofuels (capped at 5% of the overall 10% transport fuel target).

While cellulosic energy crops are favoured over first-generation biofuels by the EU, they are not as heavily favoured as biofuels from wastes, which would be quadruple-counted under the proposed changes to the RED. The key arguments in favour of wastes are that they generally have lower lifecycle GHG emissions (and hence higher GHG savings) and that they don't contribute to land use change, either directly or indirectly. The argument for high GHG savings is supported by the evidence shown in this chapter, with biofuels from agricultural wastes, forestry wastes and waste vegetable oil showing some of the highest GHG savings in Figure 2.2. However, it should be noted that a key reason for these high GHG savings is the methodology chosen by the EU for biofuels from wastes, which allocates none of the emissions related to the cultivation of these feedstocks to the biofuel, allocating them instead to the primary product (i.e. the food, fibre or timber product for which the crop was primarily grown).

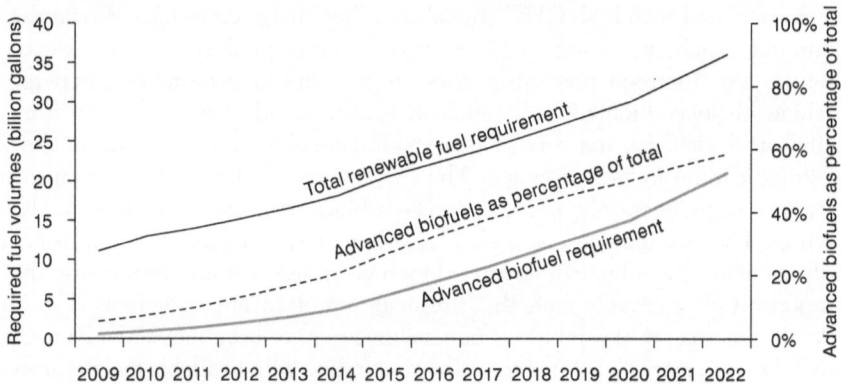

Figure 2.5 Increase in advanced biofuel requirement in the US 2009–22

Source: Environmental Protection Agency (2010)

Note: Advanced biofuels include cellulosic biofuel, biomass-based diesel and other biofuels with >50 per cent GHG savings.

A number of authors argue that the primary product allocation method used by the EU does not accurately reflect the influence that biofuel production from wastes and residues can have on the overall production system for the crop. Borrion et al. (2012) highlights three common approaches that can be used instead of a primary product allocation method, with emissions allocated according to either the relative mass, energy content or economic value of each output. The RSB provides an example of an economic value approach, allocating emissions from cultivation, transport and processing according to the relative economic value of each output (Roundtable on Sustainable Biofuels, 2012). Notably, the RSB approach does not differentiate between wastes, residues and co-products. In contrast, the EU approach draws a clear line between co-products and wastes, with emissions allocated between co-products based on energy content, but all process wastes and residues assigned zero emissions.

Drawing a clear distinction between wastes and co-products may simplify the administration of biofuel sustainability schemes, but in reality there are often shades of grey in between. It is arguable that once a waste is given some form of economic value (e.g. by converting it into a biofuel), it then has the potential to exert an influence, however small, on the overall production system. An example of this can be seen in Australia around the use of forestry and sawmill residues for bioenergy, with the Greens Party arguing that such practices could prop up forestry operations that would otherwise be uneconomic (Sydney Morning Herald, 2009). Australia's woodchip export industry has been cited as an analogous example of an industry that emerged as a way of utilising forestry wastes before becoming the primary purpose of forest harvesting in itself (Forestmedia, 2010). Thus, it is probably more accurate to view bioenergy from wastes and residues as having a lesser influence on production systems and land uses compared to energy crops, rather than having no influence at all.

Much of the motivation for favouring wastes over energy crops under the proposed changes to the RED is that they are seen as "less bad" due to their lower life cycle emissions and lesser contribution to land use change. However, as has been shown by some of the examples in this chapter, land use change should not always be seen as a negative. Energy crops have the potential to promote the revegetation of degraded land in a way that biofuels from wastes do not. Similarly, energy crops that increase the level of stored carbon in a land unit have the potential to create carbon-negative biofuels rather than simply minimising life cycle emissions. Thus, it should not be assumed that bioenergy from wastes are inherently be more sustainable than bioenergy from cellulosic energy crops.

The differing ways in which governments and NGOs have gone about setting their rules and targets for bioenergy highlight a key point about sustainability and shifting expectations. Sustainability is often presented as an absolute concept, in the sense that our ultimate aim should be to produce bioenergy that *is* sustainable. This is implied by the original name of the RSB (Roundtable on Sustainable Biofuels) and the guiding principle of the RED that "biofuel production should be sustainable" (European Parliament and Council of the European Union, 2009, p. L140/23). However, the requirements for GHG savings set by the EU, the US, the RSB and others do not really define biofuels that are conclusively "sustainable". If that were the case, why would the EU need to increase its GHG savings requirement from 35 per cent to 60 per cent, why would the US need to progressively increase the proportion of fuels that must meet a 60 per cent benchmark, and why would the RSB state that its minimum GHG requirement of 50 per cent shall increase over time?

It could be argued that only biofuels with GHG savings of 100 per cent are truly sustainable, as they would not result in any net emissions over their life cycle. Indeed, Mathews (2008) takes this argument even further and suggests that a GHG saving in excess of 100 per cent (i.e. carbon-negative) should be a prerequisite for a biofuel to be seen as sustainable. The Intergovernmental Panel on Climate Change has further illustrated this point by stating that bioenergy with carbon capture and storage (i.e. bioenergy that results in net removal of CO_2 from the atmosphere) may be an essential technology if we overshoot the target of 450 ppm CO_2 that is needed to keep global temperature rises below 2° Celsius (IPCC, 2014).

The shifting goalposts for GHG savings from bioenergy highlight that sustainability does not really represent an absolute endpoint but rather a general direction for society to head in. This view of sustainability is endorsed by authors such as Stephen Dovers, who suggests that sustainability is best viewed as a "higher order social goal" that guides us but may never be completely fulfilled, like justice, equity or democracy (Dovers, 2005, p. 7). Indeed, the RSB recognises this in the statement that its principles represent "an ever evolving standard reflecting current technical, environmental and social realities" (RSB, 2010, p. 3). As discussed in Chapter 1, it is a key aim of this book to contribute to this evolving notion of bioenergy sustainability by identifying and promoting forms of bioenergy that help us move in the direction of sustainability. This chapter has highlighted various forms of bioenergy that are capable of making significant contributions to the mitigation of climate change, as well as others that are unlikely to assist in this task due to their high levels of associated emissions. The next two chapters look at another key sustainability issue related to the production of energy crops – the role they could play in driving either land degradation or ecological enhancement.

References

Al-Riffai, P., Dimaranan, B. and Laborde, D. (2010) *Global Trade and Environmental Impact Study of the EU Biofuels Mandate*, International Food Policy Institute for the Directorate General for Trade of the European Commission, Brussels.

Augustenborg, C. A., Hepp, S., Kammann, C., Hagan, D., Schmidt, O. and Müller, C. (2012) "Biochar and earthworm effects on soil nitrous oxide and carbon dioxide emissions", *Journal of Environmental Quality*, 41: 1203–9.

BBC News (2013) "Biofuels: 'Irrational' and 'worse than fossil fuels'", www.bbc.com/news/science-environment-22127123 (accessed 16 January 2015).

Berndes, G., Bird, N. and Cowie, A. (2011) *Bioenergy, Land Use Change and Climate Change Mitigation*, IEA Bioenergy, Rotorua.

Borrion, A. L., McManus, M. C. and Hammond, G. P. (2012) "Environmental life cycle assessment of lignocellulosic conversion to ethanol: a review", *Renewable and Sustainable Energy Reviews*, 16: 4638–50.

Crutzen, P. J., Mosier, A. R., Smith, K. A. and Winiwarter, W. (2007) "N_2O release from agro-biofuel production negates global warming reduction by replacing fossil fuels", *Atmospheric Chemistry and Physics Discussions*, 7: 11,191–205.

Department for Transport (2012) *RTFO Guidance Part Two: Carbon and Sustainability Guidance*, Department for Transport (UK), London.

Dovers, S. (2005) *Environment and Sustainability Policy*, Federation Press, Sydney.

Environmental Protection Agency (2010) *Regulation of Fuels and Fuel Additives: Changes to Renewable Fuel Standard Program*, Final Rule 40 CFR Part 80 [EPA–HQ–OAR–2005–0161; FRL–9112–3] RIN 2060–A081, United States Environmental Protection Agency, Washington, DC.

Environmental Protection Agency (2013) *EPA Proposes 2014 Renewable Fuel Standards, 2015 Biomass-Based Diesel Volume*, United States Environmental Protection Agency, Washington, DC.

Ernsting, A., Rughani, D. and Boswell, A. (2007) *Agrofuels Threaten to Accelerate Global Warming*, Biofuelwatch, Edinburgh.

European Commission (2010) *Report from the Commission to the Council and the European Parliament on Sustainability Requirements for the Use of Solid and Gaseous Biomass Sources in Electricity, Heating and Cooling*, European Commission, Brussels.

European Commission (2012a) *Guidance Document: Biomass Issues in the EU ETS*, European Commission, Brussels.

European Commission (2012b) *Proposal for a Directive of the European Parliament and of the Council Amending Directive 98/70/EC Relating to the Quality of Petrol and Diesel Fuels and Amending Directive 2009/28/EC on the Promotion of the Use of Energy from Renewable Sources*, European Commission, Brussels.

European Parliament and Council of the European Union (2009) "Directive 2009/28/EC", *Official Journal of the European Union*, L140: 16–62.

Forestmedia (2010) "Don't burn forests for electricity: about biomass power", http://noforestfurnaces.org.au/about-biomass-power (accessed 13 March 2011).

Gibbs, H. K., Johnston, M., Foley, J. A., Holloway, T., ChadMonfreda, Ramankutty, N. and Zaks, D. (2008) "Carbon payback times for crop-based biofuel expansion in the tropics: the effects of changing yield and technology", *Environmental Research Letters*, 3: 1–10.

Guardian (2013) "Are biofuels worse than fossil fuels?", *The Guardian*, www.theguardian.com/environment/2013/nov/29/biofuels-worse-fossil-fuels-food-crops-greenhouse-gases (accessed 16 January 2015).

Horne, R., Grant, T. and Verghese, K. (2009) *Life Cycle Assessment: Principles, Practice and Prospects*, CSIRO Publishing, Dickson.

International Organization for Standardization (2006) *Environmental Management: Life Cycle Assessment: Principles and Framework*, ISO 14040:2006, Geneva: International Organization for Standardization.

IPCC (2014) *Climate Change 2014 Synthesis Report*, Fifth Assessment Report, Intergovernmental Panel on Climate Change, Geneva.

JRC (2014) *Solid and Gaseous Bioenergy Pathways: Input Values and GHG Emissions*, Joint Research Centre, European Commission, Brussels.

Kammann, C., Ratering, S., Eckhard, C. and Müller., C. (2012) "Biochar and hydrochar effects on greenhouse gas (carbon dioxide, nitrous oxide, methane) fluxes from soils", *Journal of Environmental Quality*, 41: 1052–66.

Lange, M. (2011) "The GHG balance of biofuels taking into account land use change", *Energy Policy*, 39: 2373–85.

Langeveld, J. W. A., Quist-Wessel, P. M. F. and Croezen, H. (2014) "Biofuel production in the Far East", in Langeveld, J. W. A., Dixon, J. and Keulen, H. v. (eds), *Biofuel Cropping Systems: Carbon, Land and Food*, Routledge, Abingdon, 159–73.

Lockwell, J., Guidi, W. and Labrecque, M. (2012) "Soil carbon sequestration potential of willows in short-rotation coppice established on abandoned farm lands", *Plant and Soil*, 360: 299–318.

Lowrance, R. and Davis, A. (2014) "Environmental sustainability of cellulosic energy cropping systems", in Karlen, D. L. (ed.), *Cellulosic Energy Cropping Systems*, Wiley, Chichester, 299–313.

Mathews, J. (2008) "Carbon-negative biofuels", *Energy Policy*, 36: 940–45.

Renewable Fuels Agency (2010) *Carbon and Sustainability Reporting Within the Renewable Transport Fuel Obligation: Technical Guidance Part One*, Office of the Renewable Fuels Agency, London.

Roundtable on Sustainable Biofuels (2012) *RSB GHG Tool Manual Version 1.0*, Roundtable on Sustainable Biofuels, Lausanne.

Roundtable on Sustainable Biomaterials (2010) *Global Principles and Criteria for Sustainable Biofuels Production: Version 2.1*, Ecole Polytechnique Federale de Lausanne, Lausanne.

Searchinger, T., Heimlich, R. A., Houghton, F., Dong, A., Elobeid, J., Fabiosa, S., Tokgoz, D. Hayes and Yu, T. (2008) "Use of US croplands for biofuels increased greenhouse gases through land-use change", *Science*, 319: 1238–40.

Subramaniam, V., May, C. Y., Muhamad, H. and Hashim, Z. (2014) "Malaysia palm oil's life cycle assessment incorporating methane capture by 2020", *Journal of Oil Palm, Environment and Health*, 5: 49–54.

Sydney Morning Herald (2009) "Blow for energy proposal", *Sydney Morning Herald*, 3 February, www.smh.com.au/news/environment/energy-smart/blow-for-energy-propos al/2009/02/02/1233423135554.html.

Vaccaria, F. P., Barontia, S., Lugatoa, E., Genesioa, L., Castaldib, S., Fornasierc, F. and Migliettaa, F. (2011) "Biochar as a strategy to sequester carbon and increase yield in durum wheat", *European Journal of Agronomy*, 34: 231–8.

van den Bos, A. and Hamelinck, C. (2014) *Greenhouse Gas Impact of Marginal Fossil Fuel Use*, Ecofys, Utrecht.

Zilberman, D., Hochman, G. and Rajagopal, D. (2010) "Indirect land use change: a second-best solution to a first-class problem", *AgBioForum*, 13: 382–90.

Chapter 3

Deforestation and land degradation

The potential risks to forests and other natural landscapes from the expansion of energy crops were highlighted in the opening paragraphs of this book. Loss of biodiversity, soil erosion, degradation of waterways and loss of stored carbon are all potential outcomes of large-scale land conversion for bioenergy crops. However, the opening paragraphs of the book were deliberately designed to be provocative, and the apocalyptic view presented therein is by no means a foregone conclusion. In practice, the impacts of energy crop expansion vary by crop and by location, as do the range of solutions available and the capacity to mitigate negative impacts in different contexts. Furthermore, while this chapter has a focus on minimising the negative impacts of energy cropping for forests and other ecosystems, it is important to remember that our focus should not just be on minimising the negatives, but also on maximising the positive contributions that energy crops can make to ecosystem health. Thus, this chapter should be read in conjunction with Chapter 4 on the potential for energy crops to contribute to ecological restoration objectives.

As discussed in Chapter 2, energy cropping can impact on ecosystem health both through direct and indirect land use change. This chapter considers indirect impacts where relevant, but is focused primarily on direct impacts, which occur when forests or other ecosystems are cleared, burned or otherwise converted for the purpose of producing energy crops. These are the impacts for which biofuel producers have the clearest responsibility, as well as the greatest capacity to bring about changes in practice.

In contrast to the direct impacts of energy cropping, biofuel producers have much influence over indirect land use change. This occurs when an energy crop displaces another crop, setting off a chain reaction that culminates in land being converted elsewhere to compensate for the production lost when the energy crop moved in. Controlling indirect impacts requires integrated responses that go beyond the bioenergy sector and cut across agriculture, forestry and other sectors that compete for land with energy cropping. Estimating how much land use change is indirectly caused by the expansion of energy crops is challenging and contentious, requiring careful analysis of global markets and land use patterns. It is also important to note that changes in global markets can bring about a variety of responses, of which clearing forest to create more arable land is only one. Other responses include increasing crop yields (e.g. through fertilisation or irrigation), multiple cropping (i.e. planting two or three crops per year), feedstock substitution (e.g. replacing palm oil with rapeseed oil), utilising biomass that is currently wasted (e.g. for animal feed) or reducing consumption of certain food and fibre products. Much of the complexity relating to land availability and market dynamics is reserved for Chapter 5 on food security, rather than being dealt with in this chapter.

Deforestation: global and regional trends

Deforestation represents one of the most significant threats to ecosystem health globally, with ramifications for biological diversity, water quality, soils and climate, as well as for the ongoing provision of food, timber, fibre and fuel. Concerns are largely centred on the tropics, with tropical forests suffering a net loss of 6.8 million hectares between 1990 and 2005, compared to small net gains in the extent of boreal, temperate and sub-tropical forests (FAO and JRC, 2012). There is some evidence that the rate of deforestation has slowed in the past decade, most notably in Brazil. For example, the most recent Global Forest Resources Assessment from the Food and Agriculture Organization of the United Nations (FAO) shows a slightly lower level of global forest conversion for 2000–2010 compared to the previous decade (FAO, 2010). However, this evidence of slowing deforestation has been contradicted by more recent satellite analysis that reveals an upwards trend for annual forest loss in the tropics across the period 2000–2012 (Hansen et al., 2013). While this satellite analysis confirms the slowdown in Brazilian deforestation, it also suggests that it has been more than offset by the increasing rates of forest loss observed for a number of other tropical countries, including Indonesia, Malaysia, Paraguay, Bolivia, Zambia and Angola.

The most significant direct driver of deforestation, estimated to account for around 80 per cent of deforestation worldwide, is agricultural expansion, which is defined broadly in this context to include commercial cropping (including energy crops), commercial grazing and subsistence agriculture (Kissinger et al., 2012). Mining, logging and urban expansion are the main drivers for the remaining 20 per cent of forest conversions. While the production of biofuel feedstocks makes up only a minor component of global agriculture, it has become a significant focus of concerns around deforestation due to three main factors. Firstly, global demand for biofuels has been rising rapidly and is forecast to continue rising. Secondly, biofuels are increasingly being produced from crops that are associated with tropical deforestation, such as soy and oil palm (Figure 3.1). Finally, the production of biofuel feedstocks is seen as a major export opportunity for tropical developing countries, many of which have a poor track record with forest protection. It has been estimated that developing countries possess 75–95 per cent of the total land that is considered available and agro-ecologically suitable for producing biofuel feedstocks (Schoneveld, 2010).

Figure 3.1 Young oil palm plantations in Malaysian Borneo, June 2010

Source: Reproduced from Wikimedia Commons (author: energie-experten.org), licensed under Creative Commons (https://creativecommons.org/licenses/by/1.0/legalcode)

Table 3.1 Competing uses for land suitable for common biofuel crops in Asia, Africa and South America

Region	Largest competing use for suitable land (forest, cultivated or other)						
	Sugar-cane	Maize	Cassava	Rape	Soya	Oil palm	Jatropha
Asia	Cultivated	Cultivated	Cultivated	Cultivated	Cultivated	Forest/ cultivated	Cultivated
Africa	Forest	Other	Forest	Other	Other	Forest/ cultivated	Forest
South America	Forest	Other	Forest	Other	Forest	Forest	Forest

Source: Data from Schoneveld (2010).

There is great variability across different regions of the world in terms of land availability and suitability for energy cropping. Table 3.1 compares Asia, Africa and South America in terms of suitable land for a range of first-generation biofuel crops. In Asia, the area of suitable land under existing cultivation exceeds the area of suitable land that is currently forested. In contrast, for Africa and South America, most suitable land is forested or under another type of natural vegetation rather than being already under cultivation. Land classed as "other" in Table 3.1 consists mainly of grassland, shrubland or sparse woodland, with Africa having the largest areas of these land types suitable for maize, cassava, soya and jatropha, and South America having the largest areas suitable for sugarcane, rape and oil palm (Schoneveld, 2010).

One interpretation of the data shown in Table 3.1 is that direct conversion of forests for energy crops is a greater threat in Africa and South America, while in Asia biofuel expansion is more likely to compete with existing crops (potentially contributing to indirect land use change elsewhere). However, oil palm is a notable exception in Asia, where much of the suitable land is currently forested. Moreover, there are a range of factors that influence land conversion other than the amount of different land types available in each region. Existing farmland may be preferred for conversion to energy crops due to its fertility, infrastructure or access to markets, while in other cases forested land may be attractive due to the potential for crop production to be supplemented by income from forest products. The level of forest protection, land tenure/ownership and foreign investment regimes also play a role, making some types of land easier to acquire for large-scale conversion than others.

While forests are the focus of much attention globally around land use change, the use of land types listed as "other" in Table 3.1 also requires close scrutiny. While such grasslands and woodlands generally do not have the same levels of carbon storage or biological diversity as tropical forests, they can provide important ecosystem functions such as watershed protection and habitat for endangered species. Moreover, while such lands are often described as "marginal", "idle" or "underutilised", they may in fact be used by local people for purposes such as shifting cultivation, intermittent grazing or firewood collection. The ecosystem health implications of converting these lands are considered in this chapter, while social impacts are dealt with in Chapters 5 and 6.

Deforestation in southeast Asia

Southeast Asia, and Indonesia in particular, has been a hotspot for concerns around forest conversion for biofuels, specifically biodiesel from palm oil. Indonesia and Malaysia dominate global production of palm oil (Figure 3.2), with Indonesia accounting for 52 per cent of global production in 2014/15 and Malaysia for 34 per cent. The expansion of palm oil plantations has been particularly rapid in Indonesia, although the expansion rate has slowed somewhat from the period 2002–7, when it averaged 16 per cent per year and Indonesia overtook Malaysia to become the world's largest producer (FAO, 2011). While biodiesel is only produced from a minority of the oil palm planted in southeast Asia, Gao et al. (2011) suggest it could be considered responsible for up to 2.8 per cent of the deforestation caused by oil palm expansion in Malaysia and up to 6.5 per cent in Indonesia.

In addition to Indonesia and Malaysia, satellite analysis by Hansen et al. (2013) also highlighted Cambodia and Laos as Asian countries with high rates of forest loss relative to their size. However, Indonesia has attracted most attention in Asia due to the extent and importance of its forests, its rapid rate of plantation expansion and its history of government policies that have encouraged forest clearance by granting concessions to a small number of companies run by

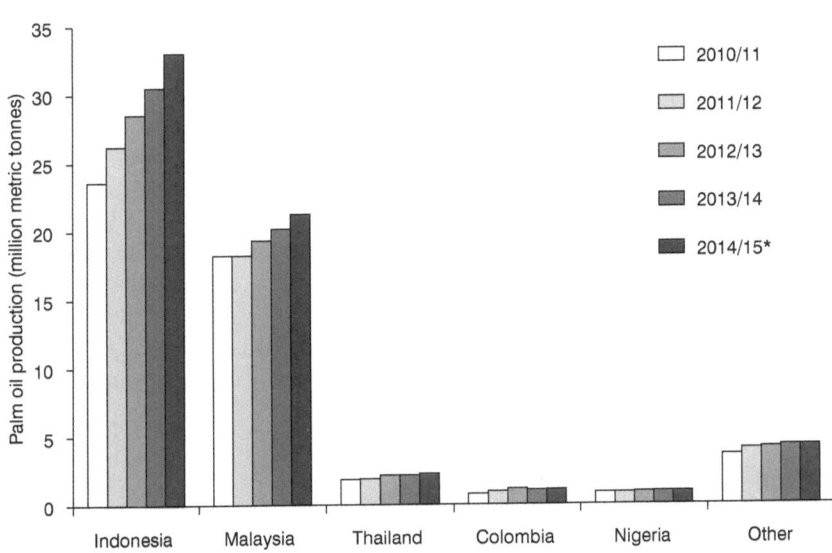

Figure 3.2 Leading producers of palm oil

Source: USDA (2015)

Note: The estimate for 2014/2015 is a projection made midway through the production year (January 2015).

influential families (Boucher et al., 2011). Indonesia's constitution grants the state considerable powers to take control of land for projects that are deemed to be in the national interest, with these powers historically used to benefit well-connected businesspeople at the expense of traditional landholders (Colchester, 2011). Recent changes have resulted in a greater emphasis on joint ventures with local communities, but this has perversely created an incentive for oil palm developers to target forested areas, where they are likely to encounter fewer landholders with which they must negotiate than in areas under existing cultivation (Sheil et al., 2009).

As discussed in Chapter 1, Indonesia introduced a moratorium on new concessions for clearing or logging environmentally sensitive areas in 2011. This moratorium covered peatlands and old-growth forest, but has been criticised for leaving large areas of forest land unprotected by classing them as "degraded", in many cases simply because they had been harvested at some point in the past. As pointed out by Edwards and Laurance (2011), many old-growth forests protected by the moratorium were in mountainous areas that were unlikely to be cleared anyway, while lowland forests providing important ecosystem functions in heavily cleared regions were left unprotected due to the fact they had once been harvested.

In a 2015 study looking at deforestation rates, concession licences and agricultural revenue, Busch et al. (2015) found evidence that the moratorium may have resulted in a lower rate of deforestation than would otherwise have occurred, but that overall deforestation rates had continued to increase and that the moratorium was inadequate to achieve Indonesia's target of a 26–41 per cent reduction in greenhouse gas emissions by 2020. They estimated that deforestation emissions during the period 2011–15 were likely to be 1.0–2.7 per cent lower than if no moratorium was in place, but that the ability of the moratorium to reduce deforestation rates was hampered by the fact that the areas it applied to had only been responsible for 15 per cent of Indonesia's emissions from deforestation during the preceding decade. Almost 60 per cent of deforestation emissions were estimated to come from areas that were not subject to any officially sanctioned concessions for clearing or logging and would therefore not be affected by a moratorium on new concessions. This finding highlights the need to enforce existing rules or introduce new controls in addition to the moratorium on new concessions.

Deforestation in Latin America

Brazil has traditionally been the biofuel powerhouse of Latin America, having pioneered the large-scale use of sugarcane for ethanol in the 1970s. Brazil continues to dominate ethanol production in Latin America and is also the largest producer of biodiesel in the region, mainly from soy. Ethanol production for 2015 is estimated at 27 billion litres, while biodiesel production is much lower at 4.4 million litres but has been growing rapidly under a government mandate that now requires biodiesel to make up 7 per cent of diesel use for transport across the country (Barros, 2014).

In addition to being the regional powerhouse for biofuel production, Brazil has also been the regional hotspot for deforestation in Latin America, with widespread global concerns emerging during the 1980s and 1990s around deforestation in the Amazon basin. However, before linking these two issues together, two important points should be kept in mind. The first is that biofuel production has played a minor role in the clearing of forests in the Amazon basin and the second is that Brazil has emerged in recent years as one of the world's success stories for slowing deforestation rates.

Sugarcane is predominantly grown in the south of Brazil, far from the frontier of Amazon deforestation. Ethanol production can contribute to indirect land use change if sugarcane displaces surrounding land uses, such as cattle grazing. However, analysis by Gao et al. (2011) for selected regions in Brazil and Columbia showed no evidence that this was occurring, with ethanol expansion instead being offset by a combination of yield increases and a reduction in the fraction of feedstock used for sugar production.

In terms of direct drivers of Amazon deforestation, the most important are beef and soy. Beef cattle are primarily raised for food and, while 20 per cent of Brazil's biodiesel production is from beef tallow, this by-product makes up only a small part of the overall economic value of raising cattle. Soybeans are the more significant feedstock, accounting for 76 per cent of Brazil's biodiesel production

(Barros, 2014). However, the 14 million Mt of soybeans forecast to be crushed for biodiesel in 2015 represent only 15 per cent of Brazil's annual soybean production (Rubio, 2015). Furthermore, biodiesel is not the only economic driver of this activity, as soy meal for animal feed is an important co-product.

Al-Riffai et al. (2010) predicted that energy crops in Brazil are likely to expand by 0.5 million ha between 2010 and 2020 in response to growing EU demand, with around 15 per cent of this expansion likely to occur at the expense of primary forest. While any loss of primary forest is likely to have negative impacts, Langeveld et al. (2013, p. 19) point out that this level of forest loss would represent a very small fraction (0.2%) of the 26 million ha of forest lost in Brazil since 2000, arguing that "the role of biofuel expansion as a major driving force for deforestation in Brazil needs to be reconsidered". In addition, there are good reasons to believe that Al-Riffai's 2010 estimate of 15 per cent of new energy cropping occurring on forested land may end up being an overestimate. This is because Brazil has dramatically reduced its deforestation rate by 70 per cent in recent years due to a combination of forest regulations, changing economic conditions and international pressure from groups such as Greenpeace (Nepstad et al., 2014).

A key element in the successful slowdown in Brazil's deforestation rate was a moratorium adopted by major soybean traders in 2006 on purchasing soybeans produced from deforested land. According to Gibbs et al. (2015), the area of soybean expansion that was associated with deforestation declined from 30 per cent in the two years prior to the moratorium to 1 per cent after its introduction. The ongoing success of the moratorium depends on the willingness of industry to maintain it and the ability of the cattle sector to continue increasing yields sufficiently to free up former grazing land for the expansion of soy. However, the success so far in decoupling soy production from deforestation in Brazil holds out hope for other regions of the world and is discussed further in the solutions section of this chapter.

Deforestation in Africa

Patterns of deforestation and land degradation in Africa share some similarities with southeast Asia and Latin America, but there are also some important differences. According to the FAO, Africa had the second-highest level of net forest loss between 2000 and 2010, with 3.4 million ha of forest being lost annually compared to 4 million in South America. Recent satellite analysis by Hansen et al. (2013) found the highest rates of overall forest loss for Africa in the Democratic Republic of Congo, Mozambique and Tanzania, while Côte d'Ivoire, Liberia and Swaziland stand out as having high rates of forest loss relative to total land area. Zambia and Angola show rapid growth in the rate of deforestation.

Compared to Latin America, cattle grazing is not as significant a driver of deforestation in Africa and logging is not as significant as in southeast Asia (Boucher et al., 2011). However, small-scale agriculture is a much more significant cause of deforestation in Africa than it is in either of the other two tropical regions. A general decline in small-scale agriculture globally has led some authors to argue that tropical deforestation worldwide is now predominantly driven by export-oriented production of agricultural commodities and timber. Fisher (2010) disputes this, arguing that Africa remains an exception to this rule, with the dominant drivers of deforestation continuing to be clearing for subsistence agriculture and extraction of timber and fuelwood to meet local needs.

Fuelwood collection has been the most significant bioenergy-related factor affecting African forests and woodlands to date. This is partly for local, non-commercial use, but also partly for conversion to charcoal for sale to nearby urban areas. Charcoal production can be a locally significant driver of deforestation and degradation around major African cities (FAO, 2007). Commercial production of biofuel feedstocks is emerging as a new land use option in Africa which could have the potential to increase deforestation. This rapid growth is apparent in a recent review of large-scale land transactions in developing countries compiled by the International Land Coalition (Anseeuw et al., 2012), which found that Africa had become the latest frontier in the global "land rush" and that biofuel production was the intended purpose for over half of the transactions they were able to verify.

The rapidly changing situation in Africa with regard to large-scale land transactions makes it difficult to determine the extent to which biofuel production has, or is likely to, impact on deforestation rates. A 2011 study by the Center for International Forestry Research (CIFOR) found that most African biofuel plantations were in areas of dry or seasonally dry forest, where deforestation is difficult to detect using satellite analysis tools (Gao et al., 2011). In addition, most biofuel plantations recorded at that time were for jatropha, which has since struggled to establish itself as a commercially viable energy crop in countries such as Tanzania, Mozambique and Mali (Romijn et al., 2014).

Deforestation in developed countries

While much of the focus on deforestation globally is on tropical developing countries, it is also important to consider developed countries such as those of the EU, the USA, Canada, Japan, Australia and New Zealand. These countries are characterised by their temperate climates, high average incomes, industrialised economies, and farming and forestry sectors that are capital-intensive with lower inputs of labour. Also, with the exception of some parts of Australia, the conversion of forests and other natural areas to agricultural land was largely complete by the end of the 20th century, with most remaining forests protected by law.

Among these nations, the US and EU dominate the production of first-generation biofuels, with the US leading the world in ethanol production and the EU the leader for biodiesel. The expansion of energy cropping in these two regions during the 1990s and 2000s has not had a significant direct impact on forested land due to large areas of surplus farmland being available and the capacity of farmers to increase the intensity of their production through inputs of fertiliser and machinery. The history of agricultural overproduction within the EU and US has played a critical role in freeing up land for the biofuel boom, as substantial areas of farmland were taken out of production during the 1980s and 90s under the EU's Common Agricultural Policy (CAP) and the US Conservation Reserve Program (CRP). The resting of this land was seen as critical to preventing overproduction, maintaining stable crop prices for farmers and preventing cost blow-outs to agricultural subsidy programmes, but it has also enabled energy cropping to expand onto these areas without requiring forest clearance or large-scale displacement of food and fibre production.

Despite the availability of surplus farmland for energy cropping in the US and the EU, both jurisdictions have implemented measures to ensure that biofuel demand does not result in further clearing of forests. As discussed in Chapter 1, the EU's Renewable Energy Directive includes sustainability criteria preventing biofuel feedstocks being sourced from primary forests and land with high conservation values. In the US, the Energy Security and Independence Act of 2007 defines renewable fuel in such a way as to exclude any crops or trees planted on land cleared after 19 December 2007. However, while the EU requires biofuel producers to prove compliance with its sustainability criteria, the US EPA has deemed all biofuel producers using domestically grown crops to be eligible without having to provide evidence or keep records. This decision is based on the EPA's assessment that existing farmland (including cropland, pastureland and CRP land) should be sufficient to meet biofuel demands in the near term and that economic factors are likely to drive production to these areas rather than uncleared land (Environmental Protection Agency, 2010). The food security implications of biofuels displacing food crops are discussed in Chapter 5.

Australia is something of an exception within the developed world, as its history of land clearing is much more recent than North America, Europe or Japan. Forests and woodlands were still being cleared at high rates in Australia as recently as the mid-2000s, especially in the states of Queensland and New South Wales (Figure 3.3). However, two factors combine to make energy cropping a low threat to the forests and woodlands of Australia. The first factor is the imposition of stronger controls on clearing in Queensland and New South Wales, which led to the dramatic decline in clearing rates that can be observed after 2006, especially with regard to first-time clearing (i.e. forests that had never been cleared before). The second factor is the low significance of energy cropping in the areas most at risk from broad-scale clearing. Most of Australia's bioenergy is sourced from wastes and residues, such as ethanol from waste wheat starch, biodiesel from waste vegetable oil and electricity from bagasse, landfill gas and wood waste. Sugarcane for ethanol is the most significant energy crop, but cane growing is restricted to coastal lowlands in the northeast that were largely cleared by the early 1900s.

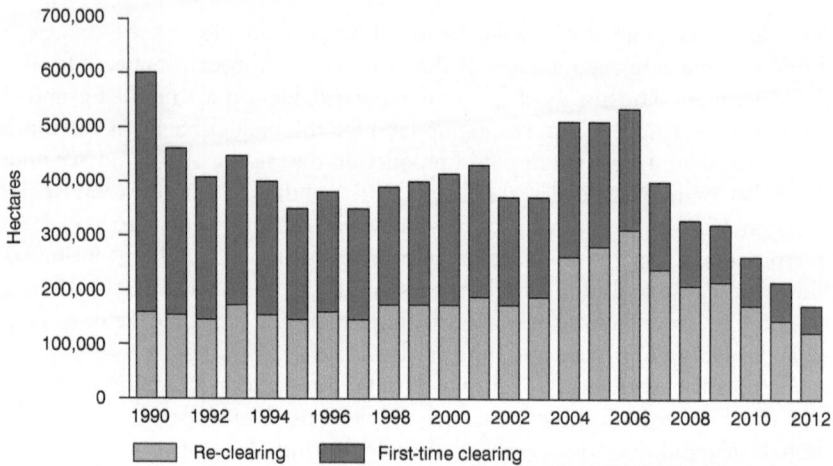

Figure 3.3 Forest clearing in Australia, first-time and re-clearing 1990–2012

Source: Reproduced from Department of the Environment (2014) under Creative Commons by Attribution 3.0 Australia (http://creativecommons.org/licenses/by/3.0/au)

While energy cropping does not pose a significant deforestation threat in Australia, it is worth noting that the use of forestry residues for bioenergy has been identified as a threat to native forests by some stakeholders following a number of proposals that emerged in the late 1990s and early 2000s (Raison, 2006). Groups such as the Wilderness Society (www.wilderness.org.au) have campaigned against the use of these "wastes" on the grounds that it could increase overall forest harvest rates, expand harvesting into old-growth areas and support harvesting that would otherwise be uneconomic. The counter-arguments put forward by the National Association of Forest Industries are that forest residues are used for bioenergy in many other developed countries and that negative impacts would be unlikely due to the low price of biomass for energy and restrictions on both the area of forest and the amount of forest material that is available for harvest (Rutovitz and Passey, 2004).

Potential solutions

The most direct way to prevent the conversion of forests and other areas with high conservation values is to introduce effective regulations in the country where the forest conversion is taking place. The decline in land clearing in Australia following the introduction of state-based native vegetation laws demonstrates how such an approach can be effective. The key advantages of land clearing regulations are that they are applied directly to the activity that causes the damage and that they can cover clearing for a range of purposes, including energy cropping, food production, timber plantations and urban development. However, there are a number of barriers to the imposition of effective land clearing laws across all global hotspots for deforestation, as highlighted by the examples cited from Indonesia and Africa. Key barriers include opposition from agribusinesses, landowners and local communities, as well as corruption and a lack of resources for enforcement.

Reform of the Brazilian Forest Code has played an important regulatory role in reducing deforestation rates in the Brazilian Amazon region. However, it also demonstrates the importance of getting local stakeholders on board with any new regulatory measures. A 1996 change to the Code increased the proportion of land that must be maintained as forest from 50 to 80 per cent. Nepstad et al. (2014) argues that this change made compliance unattainable for most landholders. This, coupled with a lack of resources for enforcement, diminished the credibility of the law. Later reforms included the introduction of an amnesty for previous illegal deforestation, the development of a rural cadastral database to aid enforcement and the creation of incentives for compliance, including restricting non-compliant landholders from accessing agricultural credit (Stickler et al., 2013).

In addition to placing broad controls on the clearing of Amazon rainforest, the Brazilian Government has also implemented agro-ecological zoning rules that restrict the expansion of certain crops. Zoning was introduced in 2009 for sugarcane expansion and in 2010 for expansion of oil palm (Leopold, 2010). Both sets of zoning rules prohibit the expansion of crops at the expense of primary forest, but a key difference exists in relation to areas already deforested. While sugarcane is prohibited from expanding into previously deforested areas of the Amazon, the zoning arrangements for oil palm actively encourage expansion into such areas on the basis that the plantations may help restore some of the ecological functions lost when the forest was originally cleared (Butler, 2011). Complementing these restrictions are incentives such as access to agricultural credit and insurance for landholders who comply with the rules.

The importance of striking a balance between regulatory restrictions and incentives for compliance is also highlighted by the example of the Indonesian moratorium on forest clearing. In that case, the incentive was provided by a $1 billion deal between Norway and Indonesia as a REDD+ initiative (reducing emissions from deforestation and forest degradation). Such initiatives have been controversial within the UN Framework Convention on Climate Change (UNFCCC), with some stakeholders resistant to the idea of paying countries not to destroy their forests, as well as holding concerns around additionality (whether the forest would have been protected anyway) and leakage (whether the clearing will simply shift elsewhere). However, others argue that such financial incentives are required to offset the opportunity cost of not clearing forest for cropping, with Stickler et al. (2013) arguing that carbon payments may be essential to maintaining the decline in deforestation rates that Brazil has managed to achieve since 2005.

A key difference between the Brazilian and Indonesian deforestation examples is the presence of strong supply chain interventions in the case of Brazil. As discussed previously, international pressure on key participants in the supply chain for Brazilian soy led to the implementation of the "Soy Moratorium" by all major buyers of Amazon soybeans. This greatly restricted the sale of soybeans grown on land that had been cleared after July 2006 and was followed up by the so-called "Cattle Agreement" among major beef processors, which imposed a barrier to the sale of cattle produced from land cleared after October 2009. There is also substantial international pressure on key participants in the palm oil supply chain, but this is yet to lead to the same level of pressure on palm oil producers in southeast Asia. While the Roundtable on Sustainable Palm Oil (RSPO) requires members to demonstrate that they are not using land cleared after 2005, as of mid-2015 only 40 per cent of the palm oil industry were RSPO members and only 18 per cent of global palm oil supplies were RSPO-certified (RSPO, 2015).

Within the biofuel sector there have also been attempts at supply chain intervention to prevent deforestation. Two examples of this were introduced in Chapter 1 – the sustainability provisions of the EU's Renewable Energy Directive (RED) and the global principles of the Roundtable on Sustainable Biomaterials (RSB). Table 3.2 compares the criteria used by the EU and RSB to two prominent standards for forestry and agriculture – those of the Forest Stewardship Council (FSC), a prominent NGO in the forestry sector, and the Sustainable Agriculture Network (SAN), which maintains the standards that are used for certification by the Rainforest Alliance.

Each of the four standards presented in Table 3.2 proscribe the clearing of forests and other high conservation value areas for the purposes of energy cropping, forestry plantations or agriculture. However, some differences are apparent in the way the standards define which areas are worthy of protection. The EU rules are highly prescriptive about the types of ecosystems that cannot be converted to biofuel production. Undisturbed forests and grasslands cannot be cleared on biodiversity grounds, while the risk to carbon stocks is cited as the reason for preventing conversion of wetland, peatland and continually forested areas. In contrast, the RSB does not list the habitat types that may qualify as having "conservation values of global, regional or local importance". This is instead left to an initial screening exercise for each biofuel project.

The FSC standard is primarily concerned with protecting forests and thus requires that no "natural forest" is converted for plantations. However, this term has been broadened considerably between the 2002 and 2012 versions of the standard to make it clear that it covers forests subject to regular harvest, secondary forests that have regrown after past clearing and certain types of woodland and savannah. Plantations may even qualify as natural forest if they develop sufficient complexity, structure and biological diversity over time.

The SAN requires broad protection of "high value ecosystems", but rather than limit itself to a set of clearly defined ecosystem types, it gives only a list of examples including primary and secondary forests, grasslands, rivers and swamps. It is also notable that the RSB, FSC and SAN standards all require management practices to protect soil health and water quality, while the EU standards focus only on which types of land may be converted.

Table 3.2 Sustainability criteria relating to deforestation and land degradation in the EU, RSB, FSC and SAN standards

EU RED 2009	RSB 2010	FSC 2012	SAN 2010
Raw materials shall not be obtained from: • land with high biodiversity value (primary forest, nature protection areas, highly biodiverse grassland) • land with high carbon stock (including wetland and continuously forested area) • peatland that has been drained	Biofuel operations shall: • avoid negative impacts on biodiversity, ecosystems, and conservation values (including no production from "no-go" zones) • implement practices that seek to reverse soil degradation and/or maintain soil health. • maintain or enhance the quality and quantity of surface and ground water resources	The organisation shall: • identify and implement effective actions to prevent negative impacts of management activities on the environmental values (and repair those that occur) • not convert natural forests to plantations, nor natural forests or plantations to any other land use (except for very limited areas if it provides conservation benefits and does not threaten high conservation values) • implement strategies and actions that maintain and/or enhance the identified high conservation values • manage infrastructural development, transport activities and silviculture so that water resources and soils are protected	• All existing natural ecosystems must be identified, protected and restored • No high value ecosystems must have been destroyed by or due to purposeful farm management activities after November 2005 • Ecosystems that provide habitats for wildlife living on the farm, or that pass through the farm during migration, must be protected and restored • The farm must have a water conservation programme that ensures the rational use of water resources • The farm must execute a soil erosion prevention and control programme

Sources: European Parliament and Council of the European Union (2009); Roundtable on Sustainable Biomaterials (2010); Forest Stewardship Council (2012); Sustainable Agriculture Network (2010)

Note: Some criteria indirectly related to deforestation have not been included (e.g. legal compliance, social impacts).

Despite the minor differences in approach, the broad similarity of the four standards shown in Table 3.2 raises the question of whether biofuel-specific standards are really required at all. Could not a single set of standards be applied to all products, including food, timber and biofuels? One argument in favour of singling out biofuels is that they are one of the fastest-growing drivers of deforestation. This view is supported by statistics showing much faster growth for biodiesel relative to other palm oil products in the EU (Gerasimchuk and Koh, 2013), as well as a greater number of large land transactions for biofuels than for food production in developing countries (Anseeuw et al., 2012). However, while energy cropping may be a fast-growing land use, it still has a long way to go to catch up with other major drivers of deforestation, namely cattle grazing, soybean cropping and oil palm for food (Kissinger et al., 2012). Biofuels do play a role in the creation of new soy and oil palm plantations, but they still account for only a minority of production, with biodiesel production accounting for only 15 per cent of soy production in Brazil (Barros, 2014) and 6.5 per cent of palm oil production in Indonesia (Gao et al., 2011).

Aside from market share and growth rates, another factor that needs to be taken into account when considering how best to apply standards is the nature of the supply chain and the types of stakeholders that are contributing to demand for biofuels. Biofuels are often less visible to the end consumer than food, timber or paper products, being blended into other vehicle fuels rather than being specifically branded as biofuels. Thus, while consumers may be familiar with wood or paper products stamped with the logo of the FSC, or food products bearing the distinctive frog logo of the Rainforest Alliance, they are less likely to encounter the logo of the RSB (Figure 3.4). However, biofuels are not really unique in this respect, as commodities like soy are often hidden from end consumers in processed products and meat from animals fed on soy meal.

One supply chain element that is unique to biofuels is the extent to which governments around the world have actively encouraged their use, through fuel-blending mandates and subsidies. It is the combination of low visibility to end users and high levels of government involvement that makes the supply chain environment for biofuels different to that for most other commodities – and creates an argument for different types of supply chain interventions to ensure

Figure 3.4 Trademarks that may be displayed on products certified by (*a*) the Rainforest Alliance, (*b*) the Roundtable on Sustainable Biomaterials and (*c*) the Forest Stewardship Council

Source: Reproduced by kind permission of the Rainforest Alliance, Forest Stewardship Council and the Roundtable on Sustainable Biomaterials

sustainability. As governments are one of the main drivers of demand for biofuels (through mandates and other policies), it stands to reason that they should have greater involvement in developing and/or endorsing sustainability standards that they do in other sectors like forestry or food production. Evidence for this can be seen in the EU's endorsement of the RSB and other standards under its Renewable Energy Directive, as well the adoption of the RSB standards by the New South Wales state government in Australia. However, while the application of sustainability standards may differ to the higher level of government involvement, this should not be seen as proof that biofuel crops represent a greater threat than other crops in terms of deforestation or require tougher restrictions. The question of whether bioenergy represents a unique threat that requires a unique response is discussed further in Chapter 8.

One final option to combat the risk of deforestation from biofuels is to preference feedstocks that are seen to pose a lower risk. This is part of the motivation behind the EU's promotion of biofuels from wastes and cellulosic feedstocks through measures that allow them to be double-counted against biofuel targets (and potentially quadruple-counted if current proposals are adopted). Biofuels from wastes or by-products that constitute very little of a crop's overall value are unlikely to drive land use change, including deforestation. Cellulosic feedstocks may drive deforestation, but this is considered less likely than for crops like oil palm or soy due to the lower value of the biomass. This could change if technological breakthroughs around advanced biofuels result in greatly increased demand for cellulosic feedstocks. However, the abundance of waste cellulose that currently exists from agriculture, forestry and industrial processes makes it unlikely that cellulosic crops would become major drivers of deforestation. These measures are discussed further in Chapter 5, as they also have implications for food security.

Conclusion

Overall, the evidence suggests that the production of biofuel feedstocks is far from being the dominant driver of deforestation worldwide. Moreover, there is no compelling reason to see them as unique and in need of tougher restrictions than are applied to other land use activities. However, deforestation remains a major threat to biodiversity, ecosystem health and local and indigenous people in many regions of the world and bioenergy producers and consumers must do their part alongside other agricultural industries to ensure that they are reducing rather than exacerbating the problem.

There is a strong argument that the different supply chain characteristics of differing commodities should be taken into account when devising strategies to counter deforestation. In the case of agricultural and forestry products that are highly visible to end users, such as coffee or office paper, it may make sense to employ certification schemes that allow consumers to select products based on their sustainability. In the case of soy and other products that are often processed or used as animal feed, the most effective point to target in the supply chain is likely to be large-scale buyers of raw produce, as shown by the effectiveness of the Soy Moratorium in Brazil. For biofuels, governments in developed countries represent key sources of demand that can be targeted to enhance protection of forests at the feedstock production end.

The various policy options discussed here will be revisited again in subsequent chapters. Direct regulation, supply chain interventions and promotion of alternative production systems are all responses with differing degrees of relevance for issues such as greenhouse gas abatement, ecological restoration, food security and land rights. The next chapter looks at ecological restoration and considers whether the producers of energy crops can go further that simply preventing the destruction of forests and other important ecosystems and instead find ways to actively enhance ecological functions and values.

References

Al-Riffai, P., Dimaranan, B. and Laborde, D. (2010) *Global Trade and Environmental Impact Study of the EU Biofuels Mandate*, International Food Policy Institute for the Directorate General for Trade of the European Commission, Brussels.

Anseeuw, W., Wily, L. A., Cotula, L. and Taylor, M. (2012) *Land Rights and the Rush for Land: Findings of the Global Commercial Pressures on Land Research Project*, International Land Coalition, Rome.

Barros, S. (2014) *Brazil: Biofuels Annual*, USDA Foreign Agricultural Service, Washington, DC.

Boucher, D., Elias, P., Lininger, K., May-Tobin, C., Roquemore, S. and Saxon, E. (2011) *The Root of the Problem: Drivers of Deforestation*, Union of Concerned Scientists, Cambridge, MA.

Busch, J., Ferretti-Gallon, K., Engelmann, J., Wright, M., Austin, K. G., Stolle, F., Turubanova, S., Potapov, P. V., Margono, B., Hansen, M. C. and Baccini, A. (2015) "Reductions in emissions from deforestation from Indonesia's moratorium on new oil palm, timber, and logging concessions", *Proceedings of the National Academy of Sciences of the United States of America*, 112: 1328–33.

Butler, R. (2011) "Could palm oil help save the Amazon?", http://news.mongabay.com/2011/0614-amazon_palm_oil.html (accessed 14 April 2015).

Colchester, M. (2011) *Palm Oil and Indigenous Peoples in South East Asia*, International Land Coalition, Rome.

Department of the Environment (2014) *National Inventory Report 2012: Volume 2*, Commonwealth of Australia, Canberra.

Edwards, D. P. and Laurance, W. F. (2011) "Loophole in forest plan for Indonesia", *Nature*, 477: 33.

Environmental Protection Agency (2010) *Regulation of Fuels and Fuel Additives: Changes to Renewable Fuel Standard Program*, Final Rule 40 CFR Part 80 [EPA–HQ–OAR–2005–0161; FRL–9112–3] RIN 2060–A081, Environmental Protection Agency, Washington, DC.

European Parliament and Council of the European Union (2009) "Directive 2009/28/EC", *Official Journal of the European Union*, L140: 16–62.

FAO (2007) *Forests and Energy in Developing Countries*, Food and Agriculture Organization of the United Nations, Rome.

FAO (2010) *Global Forest Resources Assessment 2010: Main Report*, Food and Agriculture Organization of the United Nations, Rome.

FAO (2011) *Southeast Asian Forests and Forestry to 2020: Subregional Report of the Second Asia-Pacific Forestry Sector Outlook Study*, Food and Agriculture Organization of the United Nations, Rome.

FAO and JRC (2012) *Global Forest Land-Use Change 1990-2005*, Food and Agriculture Organization of the United Nations and Joint Research Centre, Rome.

Fisher, B. (2010) "African exception to drivers of deforestation", *Nature Geoscience*, 3: 375–6.

Forest Stewardship Council (2012) *FSC International Standard: FSC Principles and Criteria for Forest Stewardship*, Forest Stewardship Council, Bonn.

Gao, Y., Skutsch, M., Masera, O. and Pacheco, P. (2011) *A Global Analysis of Deforestation Due to Biofuel Development*, Center for International Forestry Research, Bogor.

Gerasimchuk, I. and Koh, P. Y. (2013) *The EU Biofuel Policy and Palm Oil: Cutting Subsidies or Cutting Rainforest?* International Institute for Sustainable Development, Winnipeg, OH.

Gibbs, H. K., Rausch, L., Munger, J., Schelly, I., Morton, D. C., Noojipady, P., Soares-Filho, B., Barreto, P., Micol, L. and Walker, N. F. (2015) "Brazil's soy moratorium", *Science*, 23 January: 377–8.

Hansen, M. C., Potapov, P. V., Moore, R., Hancher, M., Turubanova, S. A., Tyukavina, A., Thau, D., Stehman, S. V., Goetz, S. J., Loveland, T. R., Kommareddy, A., Egorov, A., Chini, L., Justice, C. O. and Townshend, J. R. G. (2013) "High-resolution global maps of 21st-century forest cover change", *Science*, 342: 850–53

Kissinger, G., Herold, M. and Sy., V. D. (2012) *Drivers of Deforestation and Forest Degradation: A Synthesis Report for REDD+ Policymakers*, Lexeme Consulting, Vancouver.

Langeveld, J. W. A., Dixon, J., Keulen, H. v. and Quist-Wessel, P. M. F. (2013) *Analysing the effect of Biofuel Expansion on Land Use in Major Producing Countries: Evidence of Increased Multiple Cropping*, Biomass Research, Wageningen.

Leopold, A. (2010) *Agroecological Zoning in Brazil Incentivizes More Sustainable Agricultural Practices*, The Economics of Ecosystems and Biodiversity (TEEB), Geneva.

Nepstad, D., McGrath, D., Stickler, C., Alencar, A., Azevedo, A., Swette, B., Bezerra, T., DiGiano, M., Shimada, J., Motta, R. S. d., Armijo, E., Castello, L., Brando, P., Hansen, M. C., McGrath-Horn, M., Carvalho, O. and Hess, L. (2014) "Slowing Amazon deforestation through public policy and interventions in beef and soy supply chains", *Science*, 344: 1118–23.

Raison, R. J. (2006) "Opportunities and impediments to the expansion of forest bioenergy in Australia", *Biomass and Bioenergy*, 30: 1021–4.

Romijn, H., Heijnen, S., Colthoff, J. R., Jong, B. d. and Eijck, J. v. (2014) "Economic and social sustainability performance of jatropha projects: results from field surveys in Mozambique, Tanzania and Mali", *Sustainability*, 6: 6203–35.

Roundtable on Sustainable Biomaterials (2010) *Global Principles and Criteria for Sustainable Biofuels Production: Version 2.1*, Ecole Polytechnique Federale de Lausanne, Lausanne.

RSPO (2015) "Why RSPO?", www.rspo.org (accessed 1 May 2015).

Rubio, N. (2015) *Brazil: Oilseeds and Products Annual*, USDA Foreign Agricultural Service, Washington, DC.

Rutovitz, J. and Passey, R. (2004) *NSW Bioenergy Handbook*, NSW Government, Sydney.

Schoneveld, G. C. (2010) *Potential Land Use Competition from First-Generation Biofuel Expansion in Developing Countries*, Center for International Forestry Research, Bogor.

Sheil, D., Casson, A., Meijaard, E., Noordwijk, M. v., Gaskell, J., Sunderland-Groves, J., Wertz, K. and Kanninen, M. (2009) *The Impacts and Opportunities of Oil Palm in Southeast Asia: What Do We Know and What Do We Need to Know?* Center for International Forestry Research, Bogor.

Stickler, C. M., Nepstad, D. C., Azevedo, A. A. and McGrath, D. G. (2013) "Defending public interests in private lands: compliance, costs and potential environmental consequences of the Brazilian Forest Code in Mato Grosso", *Philosophical Transactions of the Royal Society B*, 368: 20120160.

Sustainable Agriculture Network (2010) *Sustainable Agriculture Standard*, Sustainable Agriculture Network, San José, Costa Rica.

USDA (2015) "Table 11: palm oil: world supply and distribution", http://apps.fas.usda.gov/psdonline (accessed 9 February 2015).

Chapter 4

Ecological restoration and enhancement

One of the key aims of this book is to identify ways of producing bioenergy that not only avoid or minimise negative impacts, such as the deforestation threats discussed in the previous chapter, but actually result in positive impacts for ecosystems and local communities. One way that energy cropping can result in positive impacts is by enhancing the health and functionality of the land on which the energy crop is grown. In some cases, energy cropping may replace a pre-existing land use activity that was contributing to land degradation or was making the land vulnerable to future degradation. In other cases, the pre-existing land use may be relatively benign, but the energy cropping system is able to offer benefits such as increased habitat for biodiversity, protection of soils against erosion and desertification, increased levels of soil carbon, mitigation of salinity impacts or improved water quality.

The energy cropping systems that are most capable of enhancing ecosystem health and functionality are typically those involving perennial trees, shrubs or grasses. Perennial crops do not require replanting each year and thus, compared to annual crops like wheat, corn or soy, they are able to develop more extensive root systems, offer better protection for soils, provide more stable habitat and reduce disturbances from tilling that can lead to soil erosion and water pollution. However, perennial trees and grasses are often incapable of producing the sugars, starches and oils that are required for the production of first-generation biofuels. Instead, they are more commonly grown for their woody or fibrous components (lignocellulose), making them more suitable for electricity generation or advanced biofuels. Notable exceptions include oil-bearing perennial crops, such as oil palm and jatropha, as well as sugarcane, which is a perennial grass that is most commonly grown as an annual crop (i.e. harvested and replanted each growing season).

This chapter highlights examples from across the world where perennial crops grown for bioenergy have produced measurable improvements in ecosystem health. There are also many examples where the full impacts of energy cropping are not yet known or where it is uncertain that bioenergy will prove to be a viable product from a perennial cropping system. A range of policy options are also discussed for dealing with the challenge posed by the mismatch between the types of biomass that can be produced from perennial crops (i.e. lignocellulose) and the types of feedstocks that currently dominate global biofuel production (i.e. oils, starches and sugars). Along the way, this chapter also deals with the more philosophical question of how to characterise the benefits for ecosystem health that may come from promoting perennial cropping systems – does this qualify as "ecological restoration" or is there a better way to describe such outcomes?

Examples of ecosystem enhancement from bioenergy cropping systems

The opening chapter of this book highlighted the example of poplar and willow crops grown for bioenergy in Europe, North America and elsewhere. This form of short rotation coppicing (SRC) represents perhaps the clearest example of energy crops that have been shown to produce a wide range of benefits for ecosystem health. The principal woody SRC species grown in Europe are willow (*Salix* spp.) and poplar (*Populus* spp.), with black locust (*Robinia pseudoacacia*) used in some areas. As discussed in Chapter 1, these cropping systems have been shown to contribute to local environmental enhancements, including an increase in soil organic matter, improved water quality and enhanced biodiversity (Simpson et al., 2009).

The application of SRC crops in Europe varies by location, with poplar principally grown in central Europe, willow most common in the north and west, and black locust suited to Mediterranean climates. Crop densities range from 8,000 to 25,000 stools per hectare and rotation lengths range from 1 to 4 years, with willow plantations tending to be denser than poplar or black locust and harvested more frequency (Maletta and Lasorella, 2014). As plantations can be viable for up to 30 years before replanting is required, there is significant opportunity for soil organic matter to accumulate through litterfall, root decay and reduced tillage. As discussed in Chapter 2, this can result in higher levels of soil carbon compared to annual cropping systems and can thus improve the life-cycle carbon balance of bioenergy from SRC crops.

With regards to soil erosion, SRC cropping systems generally have lower rates of soil loss than annual cropping systems. However, careful management is required to reduce erosion risks from bare ground exposed during the early establishment phase, such as through the use of ryegrass or other ground cover (Lowrance and Davis, 2014).

While SRC systems do not replicate natural forest ecosystems, they have been shown to provide habitat for deer, birds and bees in Europe (Simpson et al., 2009; Dimitriou et al., 2011). Impacts will inevitably vary depending on the type of land converted to SRC crops. Planting SRC crops on land previously used for annual crops will generally increase biodiversity, but it is also important to be aware that biodiversity is likely to decline if the SRC crop is established by converting forests or mature grasslands (Lowrance and Davis, 2014). Habitat opportunities can also be increased by enhancing landscape heterogeneity (e.g. multiple species and ages), strategic placement of plantations (e.g. as wildlife corridors or buffers) and careful timing of harvesting and other disturbances.

One of the other benefits commonly cited from SRC systems in Europe is the improvement to water quality from lower fertiliser requirements, filtering of runoff or wastewater, and removal of heavy metals from contaminated soils. Dimitriou et al. (2011) reviewed a range of European studies into SRC systems, finding that the establishment of such systems on agricultural land could generally be expected to result in improved groundwater quality and reduced heavy metal concentrations.

Sweden has been a particular leader in the development of SRC willow crops for bioenergy, with over 14,000 ha supplying around 1 per cent of the country's wood fuel (González-García et al., 2012), Willow crops have also strategically located to utilise wastewater and sewage. Furthermore, willow crops have also shown potential to remove heavy metals such as cadmium and zinc from contaminated soils, while maintaining reasonable levels of productivity (Van Slycken et al., 2012).

Eucalyptus trees are another important option for SRC bioenergy systems in regions with tropical, sub-tropical and Mediterranean climates. It is estimated that there are more than 20 million hectares of eucalypt plantations worldwide, with Brazil, India and China being the dominant producers (Iglesias-Trabado and Wilstermann, 2009). However, most commercial eucalypt plantations worldwide have been established for pulp, paper and timber products, with only a small minority used for commercial bioenergy.

Brazil has a history of using some of its eucalypt plantations to produce charcoal for pig iron and steel-making, with more recent trials aimed at electricity generation (Couto et al., 2011). While there has not been as much research into the impacts of eucalypt plantations on ecosystem health as there has been for SRC crops in Europe, Couto et al. (2011) argue that plantations in Brazil are valued for their soil protection attributes and the habitat they provide for wildlife. Some plantations have also been awarded carbon credits for increasing the level of sequestered carbon relative to what was there previously (FAO, 2007).

In Australia, an expansion of commercial eucalypt plantations has coincided with increasing calls for revegetation of woodlands to reverse declining biodiversity and protect soils and water quality (e.g. Vesk and Mac Nally, 2006; Cork et al., 2006). While plantations cannot replicate natural woodlands, they can enhance certain environmental values, such as habitat for bird species, especially when established on former grazing or cropping land in heavily cleared areas (e.g. Loyn et al., 2007). Bioenergy production from a range of eucalypt species has been widely investigated as an agroforestry option across low-rainfall areas of inland Australia (Bennell et al., 2009), as well as higher-rainfall areas such as the central tablelands of New South Wales (Baumber et al., 2012).

Australia's most prominent commercial development project aimed at combining bioenergy and landscape protection is the Oil Mallee Project, based in Western Australia (WA). This programme involves a variety of eucalypt species known as mallee being trialled as SRC crops to provide bioenergy and other products such as eucalyptus oil, while also helping to mitigate dryland salinity (Stucley et al., 2012). Southwest WA is a major food-producing region with significant ecological values (Mullan, 2000), but is highly susceptible to dryland salinity as a result of the extensive replacement of deep-rooted native vegetation with shallow-rooted annual grain crops (National Land and Water Resources Audit, 2001). One of the major drivers behind the oil mallee industry is to mitigate the impacts of rising saline water tables by planting strips of mallee within wheat fields (Figure 4.1). The mallee trees' high level of evapotranspiration and deep root systems prevent the saline water from rising to the surface.

A number of commercial drivers have been explored to encourage further planting of mallee in WA, with bioenergy emerging as an important option alongside eucalyptus oil (for pharmaceutical or industrial uses), activated carbon (for filtration or purification) and the sale of carbon credits from biosequestration. The 2009 Oil Mallee Industry Development Plan (URS Australia, 2009) also identifies opportunities from emerging technologies, including liquid biofuels and biochar, with aviation fuels being a more recent focus of research interest (CSIRO, 2011; Crossin, 2014). Mallee cropping options have also been explored in the eastern state of New South Wales (e.g. Abadi et al., 2006; Baumber et al., 2011), albeit to a much lesser extent than in WA.

The impact that mallee belts have on dryland salinity remains subject to much uncertainty. While Bennett et al. (2011) found that mallee belts can cause significant reductions in groundwater recharge, they forecast that the area ultimately protected from dryland salinity was likely to be no greater than the area occupied by the mallee trees themselves. This result highlights the importance of obtaining a commercial return from mallee to justify diverting

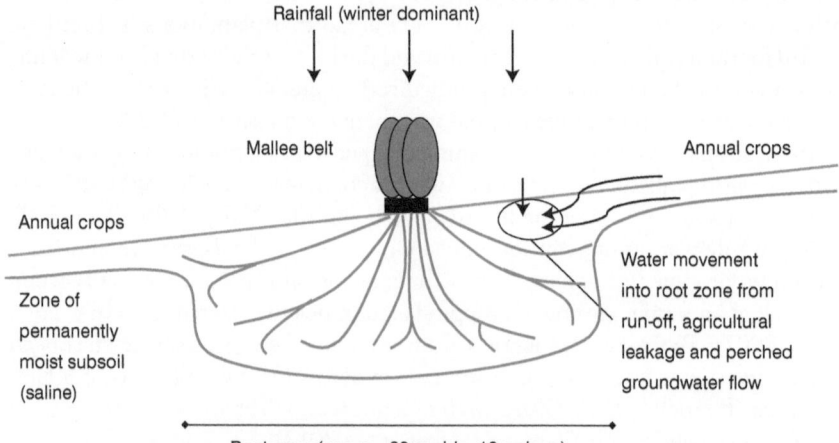

Figure 4.1 The role of mallee tree belts in mitigating dryland salinity in the West Australian wheatbelt

Source: Adapted from Yu et al. (2007)

productive land away from wheat and sheep. Aside from salinity impacts, there is also some evidence of potential benefits for biodiversity. Habitat quality in oil mallee plantations has been found to be better than adjacent crop or pasture land, but not as good as remnant native woodland (Smith, 2009).

In New South Wales, mallee plantations with mulch or grass between the rows have shown higher levels of landscape function than neighbouring sites under pasture or native vegetation using the Landscape Function Analysis (LFA) monitoring system (Baumber, 2012). LFA is a rapid assessment tool that involves dividing the landscape into patches, where resources accumulate, and inter-patches, where resources are mobile. A range of parameters such as vegetation cover, litter accumulation and evidence of erosion and deposition are assessed within each patch and inter-patch, resulting in three separate scores for soil stability, water infiltration and nutrient cycling (Tongway and Hindley, 2004). Across a variety of mallee cropping systems analysed in the central west of NSW (Figure 4.2), the highest scores for all three LFA criteria were recorded for plantings that had 3-metre row spacing, were harvested every 18 months and had been managed to ensure either grass growth or mulch cover between each row.

The plantings shown in Figure 4.2 with 3-metre row spacings produced higher scores under all three LFA criteria than the nearby native vegetation site and higher scores for infiltration and nutrient cycling than the adjacent pasture site (Baumber, 2012). However, the results of this study also highlight the importance of appropriate management practices, as the lowest LFA scores were recorded for the two older-style mallee cropping systems shown in the right-hand column of Figure 4.2. The system at the top right had been established around 25 years earlier and was deliberately designed to inhibit grass growth between rows by planting at 1.5-metre spacings and using herbicides. The system at the bottom right involved the harvest of natural stands of mallee over a 100-year period without any effort to protect soils or establish new plants.

Aside from SRC tree crops such as willows and eucalypts, bioenergy systems can also utilise perennial grasses such as miscanthus and switchgrass, with some of these systems also demonstrating positive environmental impacts on a landscape scale. Miscanthus employs the C4 metabolic pathway that is typical of tropical grasses, but is in fact most widely grown in Europe, where the hybrid *Miscanthus × giganteus* is dominant (Brancourt-Hulmel et al., 2014). Switchgrass (*Panicum virgatum*) is another C4 grass that can be grown under a variety of conditions, occurring naturally from Central America to Canada. Switchgrass is less established than miscanthus as a commercial crop and is likely to produce lower yields, in the vicinity of 10–20 tonnes of dry matter per year (t DM yr^{-1}) as opposed to more than 40 t DM yr^{-1} recorded for some miscanthus crops (Maletta and Lasorella, 2014). Other perennial grasses that have shown potential as bioenergy crops include Napier grass (*Pennisetum purpureum* Schumach.), giant reeds (*Arundo donax*) and various species of bamboo.

In terms of ecosystem health, the conversion of annual cropping systems to perennial grass crops has been shown to increase soil infiltration, improve

Figure 4.2 Mallee plantation designs assessed against pasture and native vegetation at West Wyalong, NSW, Australia. The left column shows mallee with 3-metre row spacings featuring mulch (top) and grass cover (bottom). The central column shows adjacent pasture (top) and remnant native vegetation (bottom). The right column shows older-style mallee cropping systems with 1.5-metre spacings (top) and naturally occurring mallee stands subject to harvest for over 100 years (bottom)

Source: Baumber (2012)

soil structure, sequester carbon and reduce soil erosion (Lowrance and Davis, 2014). Miscanthus assists with soil stabilisation by producing dense rhizomes that can reach depths of 2.5 metres and make up about a third of total below-ground biomass (Brancourt-Hulmel et al., 2014). Switchgrass has been targeted as potential crop for marginal land in the Central Great Plains of the US, where it could not only help to protect soils, but also provide habitat for wildlife and increase landscape heterogeneity (Hartman et al., 2011). However, management practices are critical to maximising the landscape-scale benefits of perennial grass crops, including ensuring adequate ground cover during the establishment phase, maintaining a mix of land use types to increase heterogeneity, careful management of harvesting to minimise impacts on wildlife, and ensuring that land providing critical habitat or buffering is not converted.

All of the energy crops discussed so far are grown for their lignocellulose content, making them suitable for electricity generation, industrial heat or second-generation (advanced) biofuels. However, some first-generation biofuel crops can also produce landscape benefits, depending on how they are grown. Sugarcane production (Figure 4.3) can be associated with a number of environmental problems, but in some cases can provide valuable habitat and soil protection services if managed appropriately. For example, Pearlstine and Mazzotti (2010) found that species diversity was greater in the sugarcane-dominated Everglades Agricultural Area compared with natural Everglades habitat, with contributing factors being the semi-perennial nature of sugarcane crops in this context (replanted every four years), low inputs of fertilisers and pesticides, low levels of human intervention during the growing season, strategic location adjacent to lakes and wildlife reserves, and the rotation of sugarcane with rice and other crops. Such benefits are less likely where sugarcane is harvested every year and not rotated with other crops.

In relation to soils, the replacement of annual grain crops with sugarcane has been shown to reduce negative impacts (FAO, 2008) and the Brazilian sugar industry argues that cane-growing can actually increase soil organic matter when grown on certain soil types (UNICA, 2007). However, these potential positives need to be balanced against the risks to soils if cane growers fail to maintain ground cover and practice in-field burning of trash prior to harvest.

Oil palm is another perennial crop that can be used for biofuel production. As discussed in the previous chapter, oil palm has attracted much attention globally for its role in the clearing of tropical forests and peatlands. However, some authors have argued that agroforestry based on the careful integration of oil palm into agricultural and forested landscapes could help to break the link between oil palm expansion and ecological destruction (Bhagwat and Willis, 2008). Koh et al. (2009) argue that oil palm agroforestry has the potential to offer a form of "wildlife-friendly farming" if it features low density planting with a mix of species in a mosaic of different land uses. However, these options are likely to require trade-offs against the goal of maximising the production efficiency, which generally favours large-scale monocultures with uniform characteristics.

Brazil has taken measures to promote oil palm in previously deforested areas of the Amazon basin on the basis that it could help restore some of the ecological functions lost when the forest was cleared. Incentives such as access to agricultural credit and insurance are offered to promote oil palm in previously deforested areas. Carbon sequestration and evapotranspiration are seen as the principal benefits, with prominent conservationists such as Daniel Nepstad of the Amazon Environmental Research Institute and Tim Killeen of Conservation International supporting the view that even monocultures of oil palm present a better option than the cattle ranching that has driven most of the deforestation to date (Butler, 2011).

While plantations for energy or other purposes may be able to play a role in revegetating deforested areas, it is critical that safeguards are put in place to ensure that any establishment of new plantations is not connected to the original clearing of forests. The palm oil industry in southeast Asia has made dubious claims in the past about as a tool for the revegetation of previously degraded land (e.g. US Embassy Jakarta, 2005). Such claims lack credibility due to the link between oil palm establishment and ongoing forest loss, with Sheil et al. (2009) citing evidence of fires being deliberately lit to degrade forests

Figure 4.3 Sugarcane, Dominican Republic. While commonly grown as an annual crop, sugarcane can have greater environmental value when grown in a perennial or semi-perennial manner

Source: Reproduced from Wikimedia Commons (author: I. Friviere), licenced under Creative Commons (http://creativecommons.org/licenses/by/2.5/legalcode)

and the use of palm oil licences as a back-door means of accessing new areas for logging. Hopefully, the approach taken in Brazil will be able to break this link and create a role for oil palm as part of the solution rather than the cause of the problem.

The final energy crop to be considered in this chapter is the tropical shrub jatropha (*Jatropha curcas*). Jatropha is a first-generation feedstock in the sense that it follows a well-established process (i.e. extraction of oils from its seeds for conversion to biodiesel), but it has yet to become a widespread commercial energy crop. When jatropha first rose to prominence as a potential bioenergy crop, much was made of the fact that its seeds are non-edible and it can be grown on degraded or marginal land (e.g. Francis et al., 2005; Brown, 2008; Brittaine and Lutaladio, 2010). These elements were seen as key to overcoming concerns around biofuels competing for land and water resources with food production. However, jatropha has come in for increasing criticism as its commercial development has proceeded in countries such as India, Indonesia and Nigeria (e.g. Friends of the Earth, 2010; Kant and Wu, 2011).

One key criticism of jatropha relates to its productivity on marginal land. While it can survive in dry and nutrient-poor environments, it is unlikely to produce commercially viable yields. This presents a choice between uneconomic yields on poor-quality land or competing with food production by using higher-quality land and/or irrigation. Other criticisms relate to a lack of understanding around the social and economic dimensions of growing jatropha for biodiesel, with a recent review of jatropha projects in Mozambique, Tanzania and Mali citing a range of barriers including high capital costs, long maturation times, inefficient oil-pressing technologies and strong competition from fossil fuels and palm oil (Romijn et al., 2014).

Notwithstanding the failure of many early jatropha projects and the challenges still facing its commercial cultivation, the potential for jatropha to deliver ecosystem benefits should not yet be dismissed entirely. Hedge rows of jatropha in Mali have been reported to reduce wind and water erosion, protect fields from animals and capture wind-blown soil at their base (Holthuijzen and Maximillian, 2011). Rainwater infiltration can also be enhanced by planting jatropha hedges along contours (Brittaine and Lutaladio, 2010). Even the study by Romijn et al, which identified a range of economic challenges around jatropha cultivation, highlighted that it was seen by local stakeholders to be a valuable hedge plant in areas facing the most severe environmental conditions. The key questions for jatropha are whether a production model can be found that makes bioenergy an economically viable option and whether such a model could be designed in a way so as to also provide landscape benefits. This may seem a long way off at this point in time, but it is important to remember that such a model may not need to rely on economic returns alone, provided that the combination of economic and non-economic benefits is sufficient to promote uptake in target areas.

Conceptualising bioenergy systems: ecological restoration, sustainable use or multifunctionality?

The idea of harnessing commercial drivers such as bioenergy production to enhance ecosystem health and functionality raises a number of conceptual and ethical issues. Should such activities be considered *ecological restoration*? Should the goal be to create land uses that are better than what was there before or should we be aiming to deliver the "best" ecosystem possible – whatever that might mean? What value should we place on the production of bioenergy compared to the provision of habitat or the protection of soils? How should trade-offs be managed?

In answering these questions we need to first revisit what we mean by sustainability in relation to energy crops. Chapter 2 presented the idea that sustainability is not necessarily an endpoint that can be reached, but rather a direction for society to head in. Bioenergy production systems such as SRC willow in Europe, switchgrass in the US, or oil mallee in Australia can make a strong claim to helping society move in the direction of sustainability if they are established on land that is degraded or currently being used for annual crops. Furthermore, harvesting biomass from these constructed ecosystems can create incentives for both the maintenance of such ecosystems and the establishment of more plantings. However, this is not to say that such land uses are inherently sustainable. When deciding whether or not a new land use moves us in the direction of sustainability, we need to consider what state the land was in before the change in land use, as well what other purposes the land could be used for.

Revegetating land for a combination of environmental and economic objectives can be controversial and it raises the question of whether such land use changes should be characterised as ecological restoration. Ecological restoration is a widespread land management practice that is typically aimed at conserving biodiversity, protecting soil health, maintaining water quality and both mitigating and adapting to climate change. However, perspectives vary on the ultimate goal of ecological restoration. To some, it should be aimed at restoring "naturalness" and be designed to "compensate for human influence on an ecological system in order to return the system to its historic condition" (Jordan, 1994, p. 32). To others, the very idea of naturalness is subjective and problematic. Lindenmayer et al. (2008, p. 82) argue that human perspectives will inevitably differ on what constitutes appropriate vegetation structure and condition and that, in landscapes long influenced by humans, "naturalness may not even be an appropriate characteristic to consider". Similarly, Australia's 2006 State of the Environment Report emphasises that successful restoration may require that "absolute concepts of naturalness be abandoned in favour of management for specific objectives" (Beeton et al., 2006, p. 44).

Establishing plantations for a combination of bioenergy production and ecosystem enhancement may not fit within everyone's vision of ecological restoration. However, it is important to recognise that all forms of restoration require the prioritisation of certain ecosystem attributes above others, either explicitly or implicitly. Restoration goals can range from the enhancement of one particular ecosystem attribute or function to the enhancement of multiple ecosystem attributes simultaneously.

Restoration projects also need to take into account the social and economic context in which they are taking place. Just because an energy cropping system *can* have positive impacts does not mean that it automatically *will* have such impacts in all circumstances. Finding ways to make an energy cropping system economically viable is critical to ensuring widespread uptake. However, as highlighted by the example of jatropha, there is no guarantee that the economic drivers for efficient production will align with environmental priorities to protect or restore degraded land.

For guidance on how to design and manage ecosystems that combine commercial uses with conservation objectives, it is useful to look at the extensive work that has been done on *sustainable use*, much of which has been done under the umbrella of the Convention on Biological Diversity (CBD). Sustainable use principles recognise that, in cases where certain socio-economic drivers threaten to degrade landscape health and threaten habitat for biodiversity, it may be desirable to promote alternative use activities that incentivise the maintenance of ecosystem health. The idea of deliberating promoting use activities to assist conservation is sometimes referred to as *conservation through sustainable use* (e.g. Webb, 2002; Ampt and Baumber, 2006) and is well encapsulated in the following quote from the CBD's sustainable use guidelines (CBD, 2004 p. 7): "encouraging sustainable use can provide incentives to maintain habitats and ecosystems, the species within them, and the genetic variability of the species".

As with ecological restoration, there are some questions around whether the promotion of energy cropping systems that have the potential to deliver ecosystem benefits should qualify as sustainable use. Sustainable use is sometimes viewed narrowly as relating only to "wild living resources" (e.g. IUCN, 2000), a term that could exclude replanted ecosystems. However, in other cases, sustainable use principles have been applied to the conservation of "secondary nature", which refers to environments created and managed by local people engaged in farming, forestry or other activities. One such example is the Satoyama Initiative, named after a particular landscape type in Japan that integrates managed forests, rice paddies, irrigation ponds and grasslands to provide valuable ecosystem services and crucial habitat for endangered species (Nature Conservation Bureau, 2009). Other examples of modified or replanted ecosystems in which the ongoing commercial harvest of products provides an incentive for conservation include the damar agroforests of Sumatra and the cork oak forests of the western Mediterranean.

The damar agroforests of Sumatra are so named because of the central role of the damar tree *Shorea javanica*, which provides resin for production of incense, varnish, paint, and cosmetics. The harvest of this resin provides an incentive to maintain the agroforests rather than convert them to monocultures of coffee or pepper. The Indonesian Government has granted some farmers special usage rights to damar agroforests on state land in order to buffer the World Heritage-listed Bukit Barisan Selatan National Park (Kusters et al., 2008). Similarly, the continued use of cork in wine-bottling has been championed by groups such as the World Wide Fund for Nature (WWF) in order to help maintain the 2.7 million hectares of cork oak forests in the western Mediterranean, particularly Portugal. The composition of these forests has been shaped by human management over a variety of spatial and temporal scales (Urbieta and Marañón, 2008) and their diverse mosaic habitats continue to support endangered species such as the Iberian lynx, the Iberian imperial eagle and the Barbary deer (WWF, 2006). The notion of conservation through sustainable use is encapsulated in the following argument:

> Because the forests have an economic value to local communities, people care for the forests. This helps maintain their environmental values as well as reducing the risk of fires and desertification.
>
> (WWF, 2006, p. 2)

As shown through the examples of willow, poplar and mallee eucalypts discussed previously, agroforestry systems in which bioenergy is a key output also have the potential to incentivise the protection and enhancement of biodiversity, soil health and water quality. In some cases, these systems have already been established and the ongoing production of bioenergy provides an incentive to maintain them. In others, the system is yet to be established but it could provide a beneficial alternative to current land uses that are contributing to soil or habitat loss. In time, it is possible that energy cropping systems involving willow or poplar in Europe or mallee eucalypts in Australia will come to be seen as worthy of preservation and promotion in the same way as the Mediterranean cork forests or the damar agroforests of Indonesia.

While it may be somewhat unusual to conceptualise energy cropping as a sustainable use activity, the similarities highlighted between certain energy cropping systems and other sustainable use activities suggests that some of the guidelines and principles on sustainable use developed within the CBD may have relevance to the management of energy crops. Chief among these are the Addis Ababa Principles and Guidelines for the Sustainable Use of Biodiversity. Many of these principles are quite broad and relate to overarching aspects of governance, such as the need for integrated policies and laws, international cooperation, adaptive management that incorporates multiple types of knowledge, interdisciplinary approaches to research, and effective education and communication strategies. However, a number of the principles have more specific relevance for the development of policy to promote and guide energy cropping systems with environmental benefits. These selected principles are shown in Table 4.1.

Principles 2 and 12 highlight the importance of benefit-sharing that ensures benefits and management rights flow to those who are best placed to manage the land use at a local scale. For example, if an energy cropping system offers a broad public benefit such as salinity mitigation or the provision of habitat, it is critical that benefits also flow to those managing the land, otherwise they will have no incentive to maintain the system that is providing public benefits to others. Principles 3 and 5 deal with risks of negative impacts, including the need to eliminate perverse incentives to degrade ecosystems (e.g. incentives to clear forests rather than establish crops on already cleared land). Principle 10 highlights the need to consider non-economic values, for example by designing policies that give preference to energy cropping systems with benefits for soils or biodiversity. Finally, Principle 7 relates to scale, advising that the scale of management should align with the scale of impacts (e.g. if impacts are likely to be felt across property boundaries, some form of cross-property management may be required).

Another way of thinking about land use systems that combine bioenergy production and ecosystem enhancement is through the concept of *multifunctionality*. This is a broad concept that recognises that agriculture, forestry and other land uses involve the joint production of multiple commodity and non-commodity outputs and that some of these non-commodity outputs constitute

Table 4.1 Selected principles from the Addis Ababa Principles and Guidelines for the Sustainable Use of Biodiversity

No.	Principle
2	Recognising the need for a governing framework consistent with international and national laws, local users of biodiversity components should be sufficiently empowered and supported by rights to be responsible and accountable for use of the resources concerned.
3	International, national policies, laws and regulations that distort markets which contribute to habitat degradation or otherwise generate perverse incentives that undermine conservation and sustainable use of biodiversity, should be identified and removed or mitigated.
5	Sustainable use management goals and practices should avoid or minimise adverse impacts on ecosystem services, structure and functions as well as other components of ecosystems.
7	The spatial and temporal scale of management should be compatible with the ecological and socio-economic scales of the use and its impact.
10	International, national policies should take into account: (a) Current and potential values derived from the use of biological diversity; (b) Intrinsic and other non-economic values of biological diversity; and (c) Market forces affecting the values and use.
12	The needs of indigenous and local communities who live with and are affected by the use and conservation of biological diversity, along with their contributions to its conservation and sustainable use, should be reflected in the equitable distribution of the benefits from the use of those resources.

Source: CBD (2004)

positive externalities or public goods (OECD, 2001). Food, fibre and biofuels are all examples of commodity outputs, while examples of non-commodity outputs include ecosystem services, food security and the viability of rural communities.

A multifunctional approach to framing sustainability issues across the bioenergy, plantation and restoration sectors would avoid defining any land use option in terms of a single output, but rather would recognise the joint production of multiple outputs. Even where the intent of a land manager is heavily focused around one objective (e.g. food production, biomass harvest or biodiversity conservation), there are likely to be a variety of non-target outputs such as ecosystem services or social benefits. Where land managers are actively seeking multiple outcomes, it may be inappropriate to talk about certain outcomes simply as "by-products" or "side-effects" (OECD, 2001).

Multifunctionality emphasises the joint delivery of many outputs, with individual outputs often unable to be separated from the overall system. The inseparability of many outputs is due to three main factors: technical interdependencies (e.g. erosion control through cropping patterns), non-allocable inputs (e.g. producing both biodiesel and animal feed from soy) and allocable inputs that are fixed so that the amount allocated to one output affects another (e.g. available farmland being allocated to food or to fuel).

Adding to the complexity of managing multifunctional systems is that many outputs occur at different but overlapping scales (e.g. local soil stability, regional economic resilience, national energy security). The combination of joint delivery and overlapping scales means that many outputs cannot simply be separated and addressed at the geographic scale most appropriate (e.g. global food security cannot be managed separately from local provision of ecosystem services). Rather, multiple outputs providing both private and public goods and operating at a variety of scales must be considered simultaneously.

The oil mallee example from Western Australia demonstrates the concept of multifunctionality clearly, with salinity mitigation, bioenergy production, rural economic viability and maintenance of food production being jointly delivered across multiple scales. Moreover, each of these are explicit goals rather than simply being by-products of a commodity production system. Authors such as Dornburg (2004) have advocated a shift away from viewing land uses as single-purpose systems managed for bioenergy, food or timber towards "multifunctional biomass systems" for various material and energy outputs. Other examples of multifunctional biomass systems could include shelterbelts of jatropha integrated into grazing or cropping land to provide soil protection, biodiesel feedstock and other co-products (Del Greco and Rademakers, 2006) or intercropping of soy and eucalyptus trees alongside cattle grazing in Brazil (Couto et al., 2011).

Trying to promote and manage energy cropping systems that deliver benefits for ecosystem health can create conceptual challenges because they don't necessarily fit neatly within the existing models of ecological restoration, sustainable use or multifunctionality. However, each of these conceptual models offers some important insights into how to promote and manage energy cropping in a way that maximises the potential for ecosystem enhancement. The following section looks at how policy instruments commonly employed in the restoration, plantation and bioenergy sectors can be adapted to promote bioenergy cropping systems that exhibit characteristics of ecological restoration, sustainable use or multifunctional land uses.

Policy options

A detailed consideration of policy measures relating to bioenergy, restoration and sustainability more broadly is provided in Chapters 8 and 9. The focus here is on identifying the key features that policy instruments require in order to effectively promote restoration objectives. These features fall into three main categories: incentives, metrics and expectations.

Incentives

Policy measures that are able to effectively promote ecological restoration objectives while protecting against further degradation need to feature an appropriate mix of incentives and disincentives. At the simplest level, incentives for restoration can be created through government grants or payments. Government grants are common in many developed countries, such as the National Landcare Programme in Australia or the various restoration programmes run by the Environmental Protection Agency (EPA) and the Fish and Wildlife Service (FWS) in the US. Funding programmes may also be run by non-government organisations (NGOs), with this option being very common in the US where a wide range of foundations offer grants aimed at local areas or specific habitat types. In developing countries, international NGOs such as WWF provide an important source of funding for restoration projects, along with inter-governmental agencies such as the Global Environment Facility (GEF) run by the United Nations Development Programme (UNDP).

Restoration grants are often aimed at covering some of the costs of restoration work but not generating a profit for the landholder. In other cases, a grant or payment scheme may be explicitly designed to cover the opportunity costs of taking land out of agricultural production, such as the Conservation Reserve Program (CRP) in the US. The CRP is aimed at taking highly erodible and environmentally sensitive cropland out of production and contributes to ecological restoration through reduced soil disturbance, reduced chemical use and re-establishment of grasses and trees. However, another key goal of the CRP is to support farmer incomes by simultaneously providing an alternative income source and reducing the farm production surpluses that can place downward pressure on crop prices.

The European Union's Common Agricultural Policy (CAP) is another example of a scheme developed to protect farmer incomes through subsidies and the "setting-aside" of farmland. Historically, the CAP has not had the same focus on environmental objectives as the CRP, but recent reforms have made "restoring, preserving and enhancing biodiversity" a specified aim of the CAP. This includes the exploration of new approaches, such as a pilot programme to link landholder payments directly to measurable improvements in habitat quality and biodiversity (European Commission, 2014).

Grants aimed at promoting ecological restoration represent a form of payment for ecosystem services, or PES (OECD, 2010). Other PES schemes go beyond government payments to engage the private sector in funding conservation or restoration activities. Costa Rica in particular has become well known internationally for its PES model, which has succeeded in directing voluntary payments from private companies (mostly hydroelectric plants) to landholders managing land for watershed protection, biodiversity conservation, carbon sequestration and landscape beauty (Porras et al., 2013). The demand in this case stems from a desire by corporations to be seen as socially responsible. A system of certificates for ecosystem services enables efficient over-the-counter transactions rather than having to rely on costly and time-consuming one-on-one negotiations between companies and landholders. While the main impetus behind Costa Rica's embrace of PES was a desire to slow deforestation rates (resulting in 860,000 ha of forest being protected between 1997 and 2012), the programme has also produced active reforestation on 60,000 ha and the natural regeneration of another 10,000 ha (Porras et al., 2013).

Tradable credit schemes involving carbon or biodiversity represent another form of PES. In the case of carbon trading, the ecosystem service being offered is carbon sequestration in forests, plantations or other carbon "sinks". As discussed in Chapter 2, plantations that are subject to periodic harvest can still sequester carbon relative to what was there before (e.g. cleared land), with the Carbon Farming Initiative in Australia being an example of a national scheme that has developed specific methodologies to deal with sequestration in harvested plantations.

At the international level, the United Nations Framework Convention on Climate Change (UNFCCC) provides for carbon trading across national boundaries under the Clean Development Mechanism (CDM) and Joint Implementation (JI) provisions of the Kyoto Protocol. The CDM offers the potential for investment money to flow from developed to developing countries for reforestation and afforestation projects. An example of how the CDM can be used to incentivise energy cropping is provided by the Brazilian eucalypt plantations discussed earlier in this chapter, whereby carbon credits from reforestation have been earned in the presence of periodic harvest for charcoal production. However, these plantations, operated by the Plantar Group in the state of Minas Gerais, have been somewhat controversial, with NGOs such as Carbon Market Watch (2010) questioning whether the land is eligible for reforestation and whether impacts on local communities have been fully considered. Measures to prevent negative impacts on local communities are discussed in Chapter 6.

One controversial aspect of tradable credit schemes is that the demand for ecosystem services often comes from businesses that are seeking permission to cause environmental damage in one location by offsetting their impact in another location. For example, coal-fired power plants may use the carbon offset credits they obtain by funding reforestation activities to continue burning coal. The underlying assumption that restoration in one location can adequately compensate for environmental damage in another becomes even more controversial when considering offsets for biodiversity or land degradation.

Biodiversity or habitat offsetting schemes have been employed in countries as diverse as the US, Brazil and Australia. The US Clean Water Act is a pioneering example of this approach, allowing some wetlands to be converted to other uses provided that other wetlands are created or enhanced. Brazil's Forest Code also allows landholders to use offsets to meet their requirements for retaining forested habitat (Doswald et al., 2012). The Australian state of New South Wales has implemented a tradable credit scheme covering habitat for biodiversity, known as BioBanking, which allows landholders to trade biodiversity values from their land to developers intending to impact on biodiversity through clearing elsewhere (Department of Environment and Climate Change, 2007).

Land degradation is the latest frontier in the establishment of offset markets for environmental services. Under the framework of the United Nations Convention to Combat Desertification and the Sustainable Development Goals to be introduced in late 2015, targets have emerged around "zero net land degradation" or "land degradation neutrality". This presents a potential opportunity to direct funding from activities that degrade soil fertility to those that restore them (potentially including energy cropping). However, such an approach faces many of the challenges faced by other tradable offset schemes, including ensuring the reliability of trades, defining clear quantifiable units of measure, ensuring equivalence across a wide range of land types and managing the risk of time lags or delayed benefits (Tal, 2015).

Government grants, tradable credits and other forms of payment for ecosystem services present potential opportunities to incentivise energy cropping systems that enhance ecosystem functions. However, careful consideration is required of how these measures would interact with existing policy measures around bioenergy and plantation establishment. Plantations for timber, pulpwood, food products or bioenergy are commonly promoted through incentives such as tax breaks, subsidies, land grants, investment by government-owned forestry corporations and funding for research and development. Where bioenergy is the target product, a range of other incentive programmes also come into play, including mandates for renewable electricity or biofuels, subsidies (often in the form of fuel tax exemptions) and feed-in tariffs (which require utilities to pay fixed prices for various forms of renewable energy). Some ideas for how these schemes can be designed to preference certain forms of bioenergy over others have already been presented (e.g. double-counting of fuels from certain feedstocks in the EU) and other ideas will emerge in subsequent chapters. A comprehensive analysis of how these policy options could fit together is provided in Chapter 8 and further elucidated through the case studies in Chapter 9.

Metrics

The second key element that is required for any policy measure to effectively promote multifunctional outcomes from energy cropping systems is a way of measuring such outcomes. A range of these metrics have already been developed under existing policy measures, such as the Carbon Farming Initiative and NSW BioBanking programme in the case of Australia. Under the Carbon Farming Initiative (now being rolled into the new Emissions Reduction Fund), a range of methodologies have been approved for estimating carbon sequestration from different planting systems, including farm forestry systems that are subject to periodic harvest. The methodologies employed under the Clean Development Mechanism perform a similar role on a global stage. The NSW BioBanking scheme uses a dual system of *ecosystem credits*, for general habitat gains, and *species credits*, for habitat benefits related to a specific threatened species (Department of Environment and Climate Change, 2008).

Another Australian example of a biodiversity benefit metric is that used under the BushTender programme in the state of Victoria. This programme allocates government funding to competing bids based on their predicted benefit according to the Biodiversity Benefit Index (BBI). The BBI has a maximum score of 100 per cent that takes into account the proposed management practices and the regional conservation significance of the site. The predicted gain in BBI is multiplied by the area of the proposed site to provide a predicted gain in terms of "habitat hectares" (Figure 4.4). For example, a 100 hectare site that is managed in such a way as to improve its BBI from 50 per cent to 70 per cent (i.e. a gain of 20%) would result in an overall gain of 20 habitat hectares. The metric for assessing bids is linked to a vegetation quality assessment method which is able to monitor the actual change in "habitat hectares" over time by comparing the site

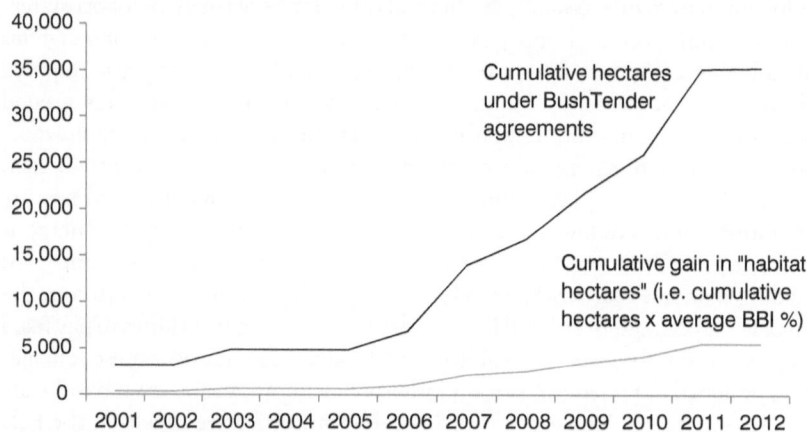

Figure 4.4 Cumulative hectares under BushTender agreements and predicted gain in habitat hectares 2001–12

Source: DEPI (2014)

to a benchmark based on a mature, long-undisturbed site of the same vegetation type, taking into factors such as landscape context and the presence of large trees, understorey plants and logs (DSE, 2004).

Apart from carbon and biodiversity, other ecosystem benefits such as erosion control, salinity mitigation and changes to water quality can be more challenging to measure and may require additional research and policy development. The CRP in the US provides an example of an integrated metric, with competing bids that provide a diverse range of benefits weighed up on a common scale known as the Environmental Benefits Index (EBI). An integrated metric such as the EBI offers the advantage of being able to compare projects with very different outcomes side-by-side, but there will inevitably be differing views on the weight that should be assigned to different ecosystem factors under such an approach. Of course, whatever metrics are chosen, there is also the challenge of predicting which forms of energy cropping systems are likely to result in which benefits and how these outcomes are affected by design feature such as species selection, monoculture vs diverse planting and harvest frequency. This need for further research is considered in Chapters 8 and 9.

Expectations

The final feature that needs to be considered for the development of policy around multifunctional energy crops is what level of expectations should be placed on them. This has an influence on how incentives and disincentives are applied and how baselines and benchmarks are set for the metrics that are used. For example, if we expect that energy crops should at least maintain conditions relative to what the land was used for previously, we might apply a negative incentive such as a fine when this expectation is not met, take no action when the expectation is met (apart from allowing the use to continue) and offer incentives to landholders who exceed these expectations.

The focus of most existing regulations aimed at plantation sustainability tends to be on maintaining rather than actively enhancing ecosystem health. However, these expectations can change over time and some recent developments around plantation sustainability have demonstrated a belief that plantations can and should enhance the functionality of the ecosystems within which they are established. Australia and New Zealand are notable examples of countries where the expectations placed on plantations have shifted over time (see Box 4.1). Such shifts in expectations may be due to "best practice" becoming progressively better, new issues arising that hadn't been previously considered, or the views of particular stakeholders becoming more prominent.

Aside from government regulations, differing expectations can also be found in the sustainability standards that have been developed by non-government stakeholders or industry participants such as retailers, processors or buyers of plantation products. In particular, such standards often reflect the expectations of consumers in developed countries, who wish to ensure that the products

Box 4.1 **Shifting attitudes towards plantations in New Zealand and Australia**

In the early 1980s, environmental NGOs were campaigning heavily against the harvesting of New Zealand's remaining native forests. One solution that was widely supported at the time was the establishment of exotic pine plantations on grazing land to provide an alternative source of timber. However, a shift in expectations occurred after most remaining forests were protected in the late 1980s, with greater scrutiny being applied to the impacts of plantations on soils, water and biodiversity (Norton, 2005). Although New Zealand pine plantations generally fare better against these criteria than the livestock grazing they replace, there have been increasing calls for plantations to approximate some of the functions of native forests, through measures such as mixed-species design, increased rotation length and better integration into landscapes.

A similar shift in attitudes can be observed among environmental NGOs and the Greens party in Western Australia. Around the turn of the millennium, such groups were vocal in their support for plantations that might be able take the pressure off native forests. This support extended as far as praising the quality of plantation woodchips and the employment opportunities they create (Tonts and Schirmer, 2005). However, a more ambivalent feeling emerged after it became apparent that plantation development was predominantly taking the form of large-scale monocultures rather than integrated farm forestry with a mix of species and plantation designs.

The latest Australian battleground is Tasmania, where the role of plantations in ending the long-running conflict over native forest harvesting elicits differing responses. The Australian Conservation Foundation (2014) endorses plantations as a key part of the forestry industry's "transition to a sustainable future", while the Wilderness Society Tasmania (2013) has opposed further plantation expansion due to issues such as chemical use and the loss of agricultural land.

These examples highlight the difficulty in selecting a single point of reference for assessing plantation sustainability. Benchmarks may be based around what plantations help prevent (the harvesting of native forests), what they have replaced (grazing or cropping) or what they could potentially become (biodiverse semi-natural forests). It is inevitable that views on what constitutes a "sustainable" plantation will vary over time and between different stakeholders.

they consume are produced in a manner that is consistent with their idea of sustainability. These standards provide an incentive to improve practices around plantation establishment and management by increasing market access (and prices in some cases) for producers who become certified under the standard.

As in the preceding chapters, four different sustainability standards have been reviewed to determine the expectations that they place on landholders. The chosen standards are those of the Forestry Stewardship Council (FSC), the Sustainable Agriculture Network (SAN), the Roundtable on Sustainable Biomaterials (RSB) and the sustainability rules for liquid biofuels under the EU's Renewable Energy Directive.

Looking across these standards, the vast majority of criteria are based on a benchmark of "maintain or enhance". In essence, this means that land managers must ensure that things don't get worse under their watch. However, there are notable criteria of the FSC and SAN standards that go beyond maintenance and require active enhancement of ecological values. The FSC requires forest managers to protect representative sample areas of native ecosystems and, if no such areas exist, to actively restore them (Forest Stewardship Council, 2012). The SAN standard goes even further, requiring plantations or farms to:

- "establish and maintain vegetation barriers between the crop and areas of human activity" (SAN, 2010, p. 20);
- "dedicate at least 30% of the farm area for conservation or recovery of the area's typical ecosystems" (SAN, 2010, p. 20); and
- "use and expand its use of vegetative ground cover to reduce erosion and improve soil fertility" (SAN, 2010, p. 42).

The inclusion of criteria around active enhancement by the FSC and SAN reinforces the idea that expectations around sustainability can shift over time and between different stakeholders. This view of sustainability may seem incongruous to some, especially those who tend to view products as either "sustainable" or "unsustainable" and look to certification bodies such as the SAN, FSC or RSB for guidance on such matters. However, it is important to remember that these organisations do not claim that their standards define a sustainable product. The FSC carefully avoids the use of the word "sustainable" at all and even organisations that do use it, such as the SAN and RSB, clearly indicate that their standards are aimed at "continual improvement" (SAN, 2010, p. 6) or represent "an ever-evolving standard reflecting current technical, environmental and social realities" (RSB, 2010, p. 3).

While at present, the focus of most efforts to define sustainable bioenergy crops tends to be on not making things worse (e.g. RSB and EU), the inclusion of some restoration outcomes under the FSC and SAN standards shows how expectations can vary over time and between different stakeholders. As more examples emerge of bioenergy crops that can actively contribute to ecosystem health, such as willow in Europe, mallee in Australia, switchgrass in the US or oil palm in Brazil, standards such as those of the EU or the RSB may shift to reflect these new expectations. There could also be a flow of expectations in the other direction, with the expectation that bioenergy crops will result in GHG savings (discussed in Chapter 2) also becoming an expectation under the FSC and SAN standards covering forestry and agricultural products (e.g. a food crop only being certified if its life-cycle GHG emissions are less than a set "best-practice" benchmark). The question of how to appropriately set expectations around these issues when designing policy measures is revisited in Chapter 9 for the case studies on Australia and Brazil.

Conclusion

In summary, it is clear that certain bioenergy cropping systems can offer the potential to enhance ecosystem health, with leading candidates being tree crops such as willow, poplar and eucalyptus, as well as perennial grasses such as miscanthus and switchgrass. However, these benefits are far from guaranteed, with sites needing to be carefully selected and management practices carefully tailored to maximise the positives while minimising the risks of negative outcomes.

In addition, there is value in expanding our thinking around energy crop sustainability to include some of the principles and guidelines that underpin the concepts of ecological restoration, sustainable use and multifunctionality. While commercial energy crops may not meet everyone's idea of ecological restoration or sustainable use, these frameworks offer insights into how we might actively promote the ecosystem benefits that some energy cropping systems could provide. The kinds of policy measures used to incentivise restoration, such as grants and tradable credits, also have relevance for energy crops that are able to provide verifiable benefits that can be replicated in new settings.

While it is not realistic to expect that all energy cropping systems will provide active ecosystem enhancement, it is important to consider how we can best promote those that do. Regulations and industry standards are key tools for ensuring that energy crops meet a basic benchmark of maintaining ecosystem values in the environments where they are established. However, other tools may be required to create incentives for energy crops that go beyond simply maintaining ecosystem health and these require careful consideration in light of the existing measures that are already used to promote bioenergy, such as grants, subsidies and mandates. These options will be revisited in Chapters 8 and 9, which explore how integrated policy measures could be used to promote energy crops that offer the greatest benefits while restricting those with the greatest risks.

References

Abadi, A., Lefroy, T., Cooper, D., Hean, R. and Davies, C. (2006) *Profitability of Agroforestry in the Medium to Low Rainfall Cropping Zone*, Rural Industries Research and Development Corporation, Canberra.

Ampt, P. and Baumber, A. (2006) "Building connections between kangaroos, commerce and conservation in the rangelands", *Australian Zoologist*, 33: 398–409.

Australian Conservation Foundation (2014) "Tasmania stands tall for ancient forests", www.acfonline.org.au/be-informed/land-forests/tasmanian-forest-protection (accessed 26 February 2014).

Baumber, A. (2012) "Harnessing bioenergy as a driver of revegetation: an analysis of policy options for the New South Wales Central West, Australia." PhD thesis, University of New South Wales, Sydney.

Baumber, A. P., Merson, J., Ampt, P. and Diesendorf, M. (2011) "The adoption of short-rotation energy cropping as a new land use option in the New South Wales Central West", *Rural Society*, 20: 266–79.

Baumber, A., Rammelt, C., Ampt, P. and Merson, J. (2012) *Bioenergy from Native Agroforestry: Planning for a Regional Industry in the NSW Central Tablelands*, Rural Industries Research and Development Corporation, Canberra.

Beeton, R. B., Buckley, K. I., Jones, G. J., Morgan, D., Reichelt, R. E. and Trewin, D. (2006) *Australia: State of the Environment 2006*, Department of the Environment and Heritage, Canberra.

Bennell, M., Hobbs, T. J. and Ellis, M. (2009) *Evaluating Agroforestry Species and Industries for Lower Rainfall Regions of Southeastern Australia: Florasearch 1A*, A report for the RIRDC/L&WA/FWPA/MDBC Joint Venture Agroforestry Program, Rural Industries Research and Development Corporation, Barton.

Bennett, D., Simons, J. and Speed, R. (2011) *Hydrological Impacts of Integrated Oil Mallee Farming Systems*, Department of Agriculture and Food, Perth.

Bhagwat, S. A. and Willis, K. J. (2008) "Agroforestry as a solution to the oil-palm debate", *Conservation Biology*, 22: 1368–9.

Brancourt-Hulmel, M., Demay, C., Rosiau, E., Ferchaud, F., Bethencourt, L., Arnoult, S., Dauchy, C., Beaudoin, N. and Boizard, H. (2014) "Miscanthus Genetics and Agronomy for Bioenergy Feedstock", in Karlen, D. L. (ed.), *Cellulosic Energy Cropping Systems*, Wiley, Chichester, 43–74.

Brittaine, R. and Lutaladio, N. (2010) *Jatropha: A Smallholder Bioenergy Crop – The Potential for Pro-Poor Development*, Food and Agriculture Organization of the United Nations, Rome.

Brown, L. R. (2008) *Plan B 3.0: Mobilizing to Save Civilization*, W. W. Norton and Company, New York.

Butler, R. (2011) "Could palm oil help save the Amazon?", http://news.monga com/2011/0614-amazon_palm_oil.html (accessed 14 April 2015).

Carbon Market Watch (2010) "Plantar–pig iron project, Brazil", http://carbonmarketw org/campaigns-issues/plantar-pig-iron-project-brazil (accessed 10 July 2015).

CBD (2004) *Addis Ababa Principles and Guidelines for the Sustainable Use of Biodi Secretariat of the Convention on Biological Diversity, Montreal.

Cork, S., Sattler, P. and Alexandra, J. (2006) *"Biodiversity" Theme Commentary Prep the 2006 Australian State of the Environment Committee*, Department of the Envir and Heritage, Canberra.

Couto, L., Nicholas, I. and Wright, L. (2011) *Short Rotation Eucalypt Plantations for Energy in Brazil*, IEA Bioenergy Secretariat, Rotorua.

Crossin, E. (2014) *Mallee Aviation Biofuels Life Cycle Assessment: Final peer-reviewed LCA report*, RMIT University, Prepared for Future Farm Industries CRC, Crawley, Western Australia.

CSIRO (2011) *Flight Path to Sustainable Aviation: Sustainable Aviation Fuel Road Map*, CSIRO, Newcastle.

Del Greco, G. V. and Rademakers, L. (2006) *The Jatropha Energy System: An Integrated Approach to Decentralized and Sustainable Energy Production at the Village Level*, Agroils, Florence.

Department of Environment and Climate Change (2007) *Biobanking: Biodiversity Banking and Offsets Scheme – Scheme Overview*, Department of Environment and Climate Change, NSW Government, Sydney.

Department of Environment and Climate Change (2008) *BioBanking Assessment Methodology* Department of Environment and Climate Change, NSW Government, Sydney.

DEPI (2014) "BushTender", www.depi.vic.gov.au/environment-and-wildlife/environmental-action/innovative-market-approaches/bushtender (accessed 21 April 2015).

Dimitriou, I., Baum, C., Baum, S., Busch, G., Schulz, U., Köhn, J., Lamersdorf, N., Leinweber, P., Aronsson, P., Weih, M., Berndes, G. and Bolte, A. (2011) *Quantifying Environmental Effects of Short Rotation Coppice (SRC) on Biodiversity, Soil and Water*, Task 43, IEA Bioenergy Secretariat, Rotorua.

Dornburg, V. (2004) *Multifunctional Biomass Systems*, Utrecht University, Utrecht.

Doswald, N., Barcellos Harris, M., Jones, M., Pilla, E. and Mulder, I. (2012) *Biodiversity Offsets: Voluntary and Compliance Regimes – A Review of Existing Schemes, Initiatives and Guidance for Financial Institutions*, UNEP-WCMC, Cambridge.

DSE (2004) *Native Vegetation: Sustaining a living landscape, Vegetation Quality Assessment Manual – Guidelines for applying the habitat hectares scoring method Version 1.3*, Department of Sustainability and Environment, State of Victoria, East Melbourne, 46p.

European Commission (2014) *Call for Proposals: Pilot On-Farm Projects to Test Result-Based Remuneration Schemes for the Enhancement of Biodiversity*, European Commission, Brussels.

FAO (2007) *Forests and Energy in Developing Countries*, Food and Agriculture Organization of the United Nations, Rome.

FAO (2008) *The State of Food and Agriculture: Biofuels – Prospects, Risks and Opportunities*, Food and Agriculture Organization of the United Nations, Rome.

Forest Stewardship Council (2012) *FSC International Standard: FSC Principles and Criteria for Forest Stewardship*, Forest Stewardship Council, Bonn.

Francis, G., Edinger, R. and Becker, K. (2005) "A concept for simultaneous wasteland reclamation, fuel production, and socio-economic development in degraded areas in India: Need, potential and perspectives of Jatropha plantations", *Natural Resources Forum*, 29: 12–24.

Friends of the Earth (2010) *Jatropha: Money Doesn't Grow on Trees*, Friends of the Earth, London.

González-García, S., Mola-Yudego, B., Dimitriou, I., Aronsson, P. and Murphy, R. (2012) "Environmental assessment of energy production based on long term commercial willow plantations in Sweden", *Science of the Total Environment*, 421–2: 210–19.

Hartman, J. C., Nippert, J. B., Orozco, R. A. and Springer, C. J. (2011) "Potential ecological impacts of switchgrass (*Panicum virgatum* L.) biofuel cultivation in the Central Great Plains, USA", *Biomass and Bioenergy*, 35: 3415–21.

Holthuijzen, W. A. and Maximillian, J. R. (2011) "Dry, hot and brutal: climate change and desertfication in the Sahel of Mali", *Journal of Sustainable Development in Africa*, 13: 245–68.

Iglesias-Trabado, G. and Wilstermann, D. (2009) "*Eucalyptus universalis*: global cultivated eucalypt forests map 2009", http://git-forestry.com/download_git_eucalyptus_map.htm (accessed 13 February 2015).

IUCN (2000) "The IUCN Policy Statement on Sustainable Use of Wild Living Resources", adopted at the IUCN World Conservation Congress, Amman, www.iucn.org/about/union/commissions/ceesp_ssc_sustainable_use_and_livelihoods_specialist_group/resources/sustainable_use_policy_statement.

Jordan, W. R. I. (1994) "Sunflower forest", in Baldwin, A. D. J., De Luce, J. and Pletsch, C. (eds), *Beyond Preservation: Restoring and Inventing Landscapes*, University of Minnesota Press, Minneapolis, MN, 17–34.

Kant, P. and Wu, S. (2011) "The extraordinary collapse of jatropha as a global biofuel", *Environmental Science and Technology*, 45: 7114–15.

Koh, L. P., Levang, P. and Ghazoul, J. (2009) "Designer landscapes for sustainable biofuels", *Trends in Ecology and Evolution*, 24: 431–38.

Kusters, K., Ruiz Perez, M., De Foresta, H., Dietz, T., Ros-Tonen, M. A. F., Belcher, B., Manalu, P., Nawir, A. A. and Wollenberg, E. (2008) "Will agroforests vanish? The case of damar agroforests in Indonesia", *Human Ecology*, 36: 357–70.

Lindenmayer, D., Hobbs, R. J., Montague-Drake, R., Alexandra, J., Bennett, A., Burgman, M., Cale, P., Calhoun, A., Cramer, V., Cullen, P., Driscoll, D., Fahrig, L., Fischer, J., Franklin, J., Haila, Y., Hunter, M., Gibbons, P., Lake, S., Luck, G., MacGregor, C., McIntyre, S., Nally, R. M., Manning, A., Miller, J., Mooney, H., Noss, R., Possingham, H., Saunders, D., Schmiegelow, F., Scott, M., Simberloff, D., Sisk, T., Tabor, G., Walker, B., Wiens, J., Woinarski, J. and Zavaleta, E. (2008) "A checklist for ecological management of landscapes for conservation", *Ecology Letters*, 11: 78–91.

Lowrance, R. and Davis, A. (2014) "Environmental sustainability of cellulosic energy cropping systems", in Karlen, D. L. (ed.) *Cellulosic Energy Cropping Systems*, Wiley, Chichester, 299–313.

Loyn, R. H., McNabb, E. G., Macak, P. and Noble, P. (2007) "Eucalypt plantations as habitat for birds on previously cleared farmland in south-eastern Australia", *Biological Conservation*, 137: 533–48.

Maletta, E. and Lasorella, M. V. (2014) "Lignocellulosic crops", in Langeveld, J. W. A., Dixon, J. and Keulen, H. v. (eds), *Biofuel Cropping Systems: Carbon, Land and Food*, Routledge, Abingdon, 208–25.

Mullan, G. (2000) "Revegetation case study: alley farming with mallee eucalypts", www.dec.wa.gov.au/programs/saving-our-species/habitat-for-nature-conservation/revegetation.html (accessed 19 May 2009).

National Land and Water Resources Audit (2001) *Australian Dryland Salinity Assessment 2000*, Land and Water Australia, Canberra.

Nature Conservation Bureau (2009) *The Satoyama Initative: A Vision for Sustainable Rural Societies in Harmony with Nature*, Nature Conservation Bureau, Ministry of the Environment, Government of Japan, Tokyo.

Norton, D. (2005) "Sustainable forest management in New Zealand", in Lindenmayer, D. and Franklin, J. (eds), *Towards Forest Sustainability*, CSIRO Publishing, Collingwood, Victoria, 167–88.

OECD (2001) *Multifunctionality: Towards an Analytical Framework*, Organisation for Economic Co-operation and Development, Paris.

OECD (2010) *Paying for Biodiversity: Enhancing the Cost-Effectiveness of Payments for Ecosystem Services*, Organisation for Economic Co-operation and Development, Paris.

Pearlstine, E. V. and Mazzotti, F. J. (2010) *Wildlife Habitat in the EAA: Farm Land, Crops, Farming Practices and Farmers Support South Florida Wildlife Species*, Institute of Food and Agricultural Sciences, University of Florida, Gainesville, FL.

Porras, I., Barton, D. N., Miranda, M. and Chacón-Cascante, A. (2013) *Learning from 20 Years of Payments for Ecosystem Services in Costa Rica*, International Institute for Environment and Development, London.

Romijn, H., Heijnen, S., Colthoff, J. R., Jong, B. d. and Eijck, J. v. (2014) "Economic and social sustainability performance of jatropha projects: results from field surveys in Mozambique, Tanzania and Mali", *Sustainability*, 6: 6203–35.

RSB (2010) *Global Principles and Criteria for Sustainable Biofuels Production: Version 2.1*, Ecole Polytechnique Federale de Lausanne, Lausanne, 29p.

SAN (2010) *Sustainable Agriculture Standard Version 3*, Sustainable Agriculture Network, San Jose, Costa Rica, 49p.

Sheil, D., Casson, A., Meijaard, E., Noordwijk, M. v., Gaskell, J., Sunderland-Groves, J., Wertz, K. and Kanninen, M. (2009) *The Impacts and Opportunities of Oil Palm in Southeast Asia: What Do We Know and What Do We Need to Know?*, Center for International Forestry Research, Bogor, Indonesia.

Simpson, J. A., Picchi, G., Gordon, A. M., Thevathasan, N. V., Stanturf, J. and Nicholas, I. (2009) *Short Rotation Crops for Bioenergy Systems: Environmental Benefits Associated with Short-Rotation Woody Crops*, Task 30, IEA Bioenergy Secretariat, Rotorua.

Smith, P. (2009) *Fact Sheet: Biodiversity Benefits of Oil Mallees*, Future Farm Industries Cooperative Research Centre, Perth.

Stucley, C., Schuck, S., Sims, R., Bland, J., Marino, B., Borowitzka, M., Abadi, A., Bartle, J., Giles, R. and Thomas, Q. (2012) *Bioenergy in Australia: Status and Opportunities*, Bioenergy Australia, Killara.

Tal, A. (2015) "The implications of environmental trading mechanisms on a future Zero Net Land Degradation protocol", *Journal of Arid Environments*, 112, Part A: 25–32.

Tongway, D. and Hindley, N. L. (2004) *Landscape Function Analysis: Procedures for Monitoring and Assessing Landscapes*, CSIRO Sustainable Ecosystems, Canberra.

Tonts, M. and Schirmer, J. (2005) "Managing social conflict in the tree plantation industry: growing consensus or deepening divisions?", in Cryle, D. and Hillier, J. (eds), *Consent and Consensus: Politics, Media and Governance in Twentieth Century Australia*, API Network, Perth, 275–96.

UNICA (2007) *Sugar Cane's Energy: Twelve Studies on Brazilian Sugar Cane Agribusiness and Its Sustainability*, UNICA – São Paulo Sugar Cane Agroindustry Union, São Paulo.

Urbieta, I. R. and Marañón, M. A. Z. T. (2008) "Human and non-human determinants of forest composition in southern Spain: evidence of shifts towards cork oak dominance as a result of management over the past century", *Journal of Biogeography*, 35: 1688–1700.

URS Australia (2009) *Oil mallee industry development plan for Western Australia*, URS Australia Pty Ltd for Oil Mallee Association of Western Australia Inc (OMA) and the Forest Products Commission (FPC), Perth.

US Embassy Jakarta (2005) "Indonesia: palm oil production a mainstay of North Sumatra economy", www.usembassyjakarta.org/econ/Sumatera_palm_oil_dec05.html (accessed 4 September 2009).

Van Slycken, S., Witters, N., Meiresonne, L., Meers, E., Ruttens, A., Van Peteghem, P., Weyens, N., Tack, F. M. G. and Vangronsveld, J. (2012) "Field evaluation of willow under short rotation coppice for phytomanagement of metal-polluted agricultural soils", *International Journal of Phytoremediation*, 15: 677–89.

Vesk, P. A. and Mac Nally, R. (2006) "The clock is ticking: revegetation and habitat for birds and arboreal mammals in rural landscapes of southern Australia", *Agriculture, Ecosystems and Environment*, 112: 356–66.

Webb, G. (2002) "Conservation and sustainable use of wildlife: an evolving concept", *Pacific Conservation Biology*, 8: 12–26.

Wilderness Society Tasmania (2013) "Statement of principles a great start – but there's more to do", www.wilderness.org.au/regions/tasmania/statement-of-principle-a-great-start-but-theres-more-to-do/?searchterm=%20plantations%20tasmania (accessed 26 February 2013).

WWF (2006) *Cork screwed? Environmental and Economic Impacts of the Cork Stoppers Market*, WWF/MEDPO, Rome.

Yu, Y., Bartle, J. and Wu, H. (2007) "Modelling mallee biomass supply in Western Australia", Bioenergy Australia Annual Conference, Curtin University of Technology and Department of Environment and Conservation, Gold Coast.

Part III

Socio-economic dimensions of energy cropping

Chapter 5

Food security

In addition to the environmental concerns surrounding climate change and deforestation that have made energy crops controversial, social issues have also risen to prominence in relation to bioenergy sustainability. Chief among these concerns is the issue of food security. Sometimes this issue is reduced a simplistic notion of "food versus fuel" – the idea that using land, water and other resources for bioenergy will inevitably reduce the resources available for producing food. The diversion of resources into energy cropping has been cited as a key factor behind rising global food prices by a number of authors, especially in the wake of the rapid price rises leading up to 2008 (e.g. Oxfam International, 2007; Dehue et al., 2007; Monbiot, 2007; Eide, 2008; Brown, 2008). However, as discussed in Chapter 1, there are a range of other factors that influence global food prices and the production of food, fuel, feed, fodder and fibre is not necessarily a zero-sum game. This chapter explores the concerns that have arisen, the research that has been undertaken to better understand the problem and the range of solutions that have been either implemented or proposed to protect food security while enabling growth in bioenergy production.

Perspectives on food security and the impact of energy cropping vary considerably. As discussed in Chapter 1, the UN Special Rapporteur on the right to food, Jean Zeigler, famously described the diversion of food crops into biofuels in 2008 as a "crime against humanity". Conversely, Calle et al. (2015) argue that adequate land exists for both food and fuel and that the primary problem is a lack of purchasing power among the world's poorer citizens. The view presented by UN Energy (a knowledge network on energy matters within the United Nations) is that bioenergy production can affect food security in a variety of ways, but that the notion of "food versus fuel" is "overly simplistic and fails to reflect the full complexity of factors that determine food security at any given place and time" (UN Energy, 2007, p. 31).

A recent report to the EU found that bioenergy has the potential to either increase or decrease food security depending on the policies behind its development and the characteristics of the local agriculture sector (AETS, 2013). EU biofuel demand was found to be a significant driver of large-scale land acquisitions in sub-Saharan Africa, but was not found to be the major cause of high food prices. Factors such as high growth in demand for food, slowing rates of productivity growth in food production, high fossil fuel prices and unfavourable weather conditions in key production areas were found to be more significant than the growth in biofuel demand. However, in the context of these wider pressures on food prices, growth in biofuel demand from first-generation feedstocks does have the potential to exacerbate food insecurity among consumers with the least purchasing power.

Rising demand for bioenergy has the potential to increase food insecurity through global-scale processes or through impacts occurring on a much more local level. At the global scale, the increased demand for biofuels may stimulate prices for crops that are suitable for either food or bioenergy feedstocks, causing large volumes to be diverted from food markets to fuel markets and making these crops unaffordable to many poorer consumers. This mechanism is dependent on the interconnectedness of international commodity markets and is most significant for first-generation biofuels produced from common agricultural crops such as corn, wheat or soy. The degree of interchangeability between crops is also important, with farmers as well as key consumers (e.g. feedlot operators) often able to switch crops depending on prevailing prices.

In addition to its global dimensions, food insecurity can also operate at a much more local level. The diversion of land away from food production can have significant impacts on local food availability and livelihoods for local communities, especially in less developed countries where communities may not be able to access global food markets due to lack of infrastructure, trade barriers or a lack of purchasing power. Such impacts are often related to large-scale land acquisitions by non-local stakeholders (German and Schoneveld, 2011), which highlights the interconnectedness between local and global dimensions of food insecurity. Price rises for key agricultural crops may stimulate land acquisitions and lead to the replacement of one crop with another. The crops that have been replaced may then rise in price due their reduced availability, potentially setting off another round of land acquisitions and land use changes.

Many of the energy crops involved in large-scale land acquisitions are common food crops, such as oil palm or sugarcane, but it is important to remember that it is not only edible crops that can impact on food security. The expansion of jatropha plantations, discussed in Chapter 4, provides an example of an inedible bioenergy crop creating concerns around food security, with land having been acquired for jatropha plantations at the expense of local food production in places ranging from Zambia to the Philippines (Anseeuw et al., 2012).

Apart from differences in geographical scale, food security impacts can also vary in terms of the speed at which they occur and how different stakeholders are affected. Diverting existing crops from food markets to biofuel markets (e.g. selling corn to an ethanol plant rather than a food processing plant) is able to occur quickly because it does not require any change in land use or production techniques (although new processing facilities and infrastructure may be required). In contrast, a major change in land use (e.g. from extensive grazing to intensive cropping) may be much slower due to the need for the land to be prepared, new infrastructure to be built, logistics networks to set up and a labour force to be engaged. This may take even longer if a change in land tenure is involved.

The effects on different stakeholders can also vary depending on the nature of the shift from food production to energy cropping. Where a farmer switches to energy crops due to their higher value without any change in land tenure, the impact on the farmer is likely to be positive, while any negative impacts of higher food prices will be felt by poor consumers. The farmers who are most likely to benefit in this way are those who are well-connected to global markets, have secure land tenure and are able to access capital that allows them to respond to rising prices by investing in new production. Consumers who suffer the most are those without land on which to grow their own crops and without the purchasing power to compete with biofuel producers for basic agricultural commodities. For example, it has been estimated by the Food and Agriculture Organization of the United Nations (FAO) that the rapid rise in global food prices leading up to 2008 resulted in an additional one hundred million people worldwide becoming food insecure (Eide, 2008).

While some farmers may benefit from the opportunity to sell their crops into biofuel markets, in other cases it is farmers themselves who can suffer from the introduction of new cash crops (including energy crops). This is particularly the case for large-scale land acquisitions where existing land tenure is insecure, and where the interests of local landholders and communities are overlooked due to poor governance, corruption and/or a lack of negotiating power (Anseeuw et al., 2012). If the change in land use is for the purposes of export, there may also be flow-on effects to local communities who relied on the previous landholders to supply local markets. In such cases, food security may decline not only for the landholders who have lost their land, but also for the local community, which has lost an affordable source of food. These issues overlap considerably with the issues of land rights and community impacts discussed in Chapter 6.

The preceding paragraphs present a rather bleak vision of the risks that energy crops can pose to food security. However, this is not the only future pathway that is possible for food and bioenergy. Many researchers have argued that sufficient land and other resources exist for the production of both food and fuel, without causing excessively high food prices for poor consumers. Furthermore, it is also possible that investment in energy cropping could actually increase the global food supply by providing new agricultural infrastructure and increasing the purchasing power of rural households in developing countries. In terms of local food security, a number of options have been proposed to ensure that existing landholders are among the beneficiaries of any land transactions and that developers of new energy cropping projects are required to maintain or enhance local food security. These issues are explored in more detail in the following sections on land use availability and policy responses.

Land use availability

A key question underpinning the "food versus fuel" debate is whether there is sufficient arable land to satisfy global demands for food while also providing feedstocks for biofuels. A number of land-use modelling studies have looked at this question in recent years, with differing results depending on the assumptions used regarding food demand, crop yields and the types of land that may be suitable for food and fuel production. A 2008 study by Veronika Dornburg and others provides a useful overview of the global modelling studies undertaken up to that point in time that looked at future land availability for food and fuel. The studies reviewed by Dornburg et al. (2008) produced a wide range of estimates for potential production of bioenergy by 2050, with 300 to 800 exajoules (EJ) per year emerging as the "medium" range of estimates from previous studies (1 EJ = 10^{18} joules). Dornburg et al. further refined this range to 200–500 EJ per year based on their own modelling, with a key assumption being that demand for food is met before any agricultural land is diverted to bioenergy use.

While the wide range of estimates highlights the uncertain nature of forecasting production levels four decades into the future, it is important to note that even the lower end of Dornburg's range would represent a substantial increase on current levels of bioenergy production. Given that current global bioenergy production from all sources (i.e. wastes, energy crops and other sources) is around 50 EJ yr^{-1} (Bauen et al., 2009), an increase to between 200 and 500 EJ yr^{-1} by 2050 would represent growth of between 300 and 900 per cent over 40 years.

Figure 5.1 shows the breakdown of bioenergy sources for Dornburg's estimate of 200–500 EJ yr^{-1} (Dornburg et al., 2010). These include residues (i.e. from agriculture and other industries), surplus forestry, energy crops from surplus agricultural land, energy crops from marginal and degraded land, and increases in productivity resulting from learning around agricultural technology. Overall, their estimated contribution from energy crops by 2050 ranged from 120 to 330 EJ yr^{-1}, depending on the productivity gains achieved and the extent to which land with water scarcity and degradation issues could be used for energy cropping. These levels are similar to those reported by Hoogwijk et al. (2009), who estimated that energy crops could provide between 130 and 270 EJ yr^{-1} by 2050 using only land that was not required to meet food, fodder, forestry or biodiversity conservation requirements.

While the results from the Dornburg and Hoogwijk studies suggest that large expansions of energy cropping are possible without jeopardising food security, other studies have come to more pessimistic conclusions. A 2009 study commissioned by OFID (OPEC Fund for International Development) found that achieving a biofuel target of 10 per cent of all transport fuels globally by 2030 would cause food prices to be 35 per cent higher than under the reference scenario, in which biofuel feedstock demand remained steady at 2008 levels (Fischer et al., 2009). Such price rises were estimated to increase the number of people at risk from hunger by 15 per cent compared to 2008 levels.

   The potential solutions proposed by the OFID study included limiting further increases in the production of first-generation biofuels to feedstocks that were demonstrably surplus to food needs and focusing on second-generation feedstocks grown on land that is not required for food or animal feed. These proposals highlight a couple of key points around land use modelling and policy development. First, the fact that sufficient arable land exists to meet both food and fuel needs does not automatically mean that food needs will be met first before any land is used for bioenergy. Land may well be used for energy crops in preference to food crops if the consumers demanding biofuels have greater purchasing power than those demanding basic foodstuffs. The second point highlighted by the OFID study is the critical role that will need to be played by technological advancements around second-generation biofuels and agricultural productivity. Policy measures have a vital role to play in addressing both these issues.

   Models used to predict changes in land use need to consider three main ways in which demand for energy crops can affect land use patterns. First, increased demand for bioenergy feedstocks may cause new areas to be brought into production. Second, land and other resources may be reallocated to produce crops that have become more profitable due to the increased demand. Third, increased crop prices may create an incentive to increase the level of productivity per unit of land, through actions such as increased fertiliser use, irrigation or growing more than one crop per year (multiple cropping). These three factors

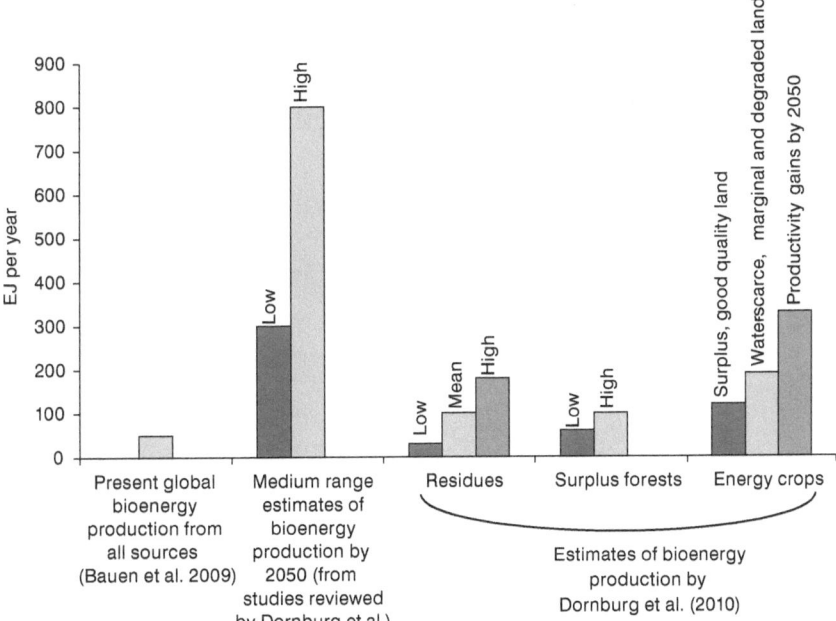

Figure 5.1 Estimates of bioenergy potentials in 2050
Sources: Dornburg et al. (2008, 2010); Bauen et al. (2009)

are critical to modelling not only food security, but also the other impacts from indirect land use change discussed in preceding chapters, such as biodiversity loss and greenhouse gas emissions from forest clearing.

With regards to bringing new land under cultivation, the greatest potential for further expansion is in Africa and South America, where just seven countries are estimated to account for 70 per cent of the potential to expand cultivated areas globally (Fischer et al., 2009). A 2010 study concluded that sufficient suitable land was theoretically "available" (i.e. not forested or currently under cultivation) in selected countries in Latin America and Sub-Saharan Africa to meet 10 per cent of domestic transport needs using first-generation biofuels, as well as allow some export of feedstocks to global markets (Schoneveld, 2010). However, the results were different for the Asian countries included in the study, where the amount of land required to supply 10 per cent of transport fuel would exceed the amount of land classed as suitable and available. Moreover, careful management of any expansion of biofuel crops would be required to ensure that they did not end up on land that is currently forested or under cultivation. Even for land classed as "available", controls are needed to ensure that any land acquisition is undertaken fairly and negative impacts are avoided for any people who may have been using the land for purposes such as fuelwood collection, shifting cultivation or as drought reserves for livestock grazing (these issues are discussed further in Chapter 6).

Regarding the reallocation of existing farmland from one crop to another, careful analysis of a variety of factors is required. As an example, the increased demand for corn in the US due to the expansion of ethanol production between 2001 and 2011 clearly contributed to a rise in the area of land planted to corn (Figure 5.2). Furthermore, US agricultural data indicates that the increase in the

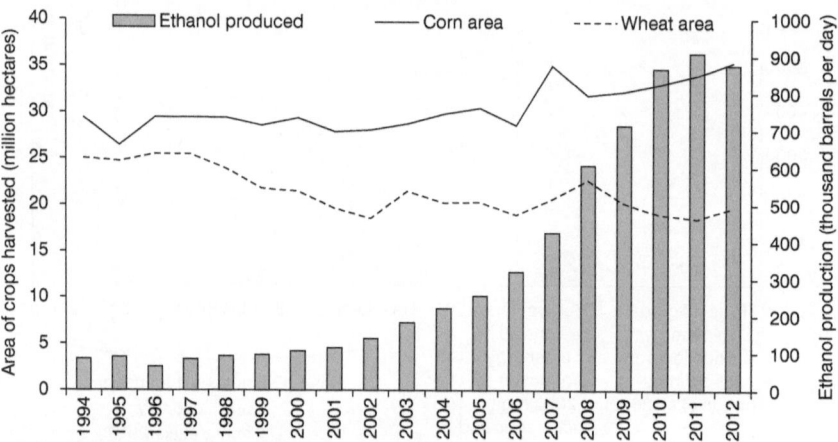

Figure 5.2 **Area of corn and wheat harvested and ethanol produced in the United States 1994–2012**

Sources: USDA (2015); Energy Information Administration (2015)

area planted to corn may have come partly at the expense of the area planted to wheat. However, determining the extent to which this may have occurred is difficult, as the land types used for each crop are not perfectly substitutable, the trend of declining wheat acreage preceded the rapid jump in ethanol demand and there are many other factors that may have contributed to in the declining use of land for wheat, such as the expansion of soybean production to supply food and animal feed markets (USDA, 2010).

Kim and Dale (2011) attempted to determine from historical data whether the increase in US biofuel production from 2002 to 2007 had impacted on land use patterns in the US and driven indirect land use change (iLUC) around the world. Focusing on the US, the only statistically significant correlation they could find between the expanding acreage of biofuel crops and a decline in other crops was in relation to cotton, which is not a food crop. They also found no evidence that growing demand for biofuels in the US had induced indirect land use change in countries that were major trading partners, which contradicts the assumptions used in a number of other studies (such as those looking at GHG impacts of iLUC discussed in Chapter 2). One possible explanation for this is that the increased demand for biofuel feedstocks had been met through productivity increases on existing farmland.

Kim and Dale are not the only researchers to argue that the historical record does not support the arguments made by authors such as Mitchell (2008) and Eide (2008) at the height of the 2008 food price peak that biofuel growth was a key cause of food price rises. Analysis by Ajanovic (2011) and Zilberman et al. (2013) failed to find clear evidence that rising biofuel demand has had a major impact on food prices. Both studies highlighted other factors that have led to higher and more volatile global food prices in recent years, such as higher oil prices, economic growth in developing countries and declining inventories of grains and oilseeds, but could not identify a significant impact from biofuel demand.

Looking at the historical record of land use change between 2000 and 2010, Langeveld et al. (2014b) found that productivity increases, particularly through multiple cropping, have been critical in helping to meet the growing demand for food and fuel. The focus of their study was land use change between 2000 and 2010 across seven countries plus the EU, which together accounted for 97 per cent of ethanol production and 77 per cent of biodiesel production in 2010. While their analysis showed an increase in land used for biofuel cropping of 25 million ha across the study area between 2000 and 2010, this did not mean that there was a decline in the availability of land for food production (Figure 5.3). On the contrary, the increase in multiple cropping (i.e. producing more than one crop per year from a unit of land) resulted in a net increase of 19 million ha in the land area available for food, feed and fibre, even after accounting for the diversion of land into biofuel crops.

Unpacking the figures from Langeveld et al. (2014b) shows that, firstly, the increase in land "dedicated" to biofuels is really only 13.5 million ha rather than 25 million ha, as a portion of the land area should be assigned to co-products such as animal feed that are produced in conjunction with biofuels. An increase of 13.5 million ha in land used for biofuels is still significant and could potentially be a cause for concern, especially as the total area of agricultural land across the study area fell by 9 million ha over the same period due to factors such as urban encroachment, forest development and land abandonment. However, this is dwarfed by the impact of increased multiple cropping across the study area, which was estimated to have freed up the equivalent of 41.5 million hectares of agricultural land. After subtracting the 13.5 million ha that shifted into biofuel production and the 9 million ha that dropped out of production entirely, the net increase in harvested area for food, feed and fibre across the study area is 19 million ha.

Overall, it is difficult to ascertain which models of land use availability represent the most realistic vision of the future. Clearly there is potential for new agricultural land to be brought under production in some parts of the world and for increases in productivity to be achieved through multiple cropping and other measures. However, it is always challenging to predict the extent to which the trends of the past can be replicated in the future. Furthermore, the fact that sufficient land exists for both food and fuel does not in itself mean that adequate land will be set aside for food and biodiversity conservation before energy crops are allowed to expand. Nor does it mean that vulnerable people will be protected from land grabs or rises in food prices. These are issues that are broader than the biofuel sector alone and require a holistic approach across all land use sectors. The various policy measures that have been proposed or implemented to address these issues are considered in the following section.

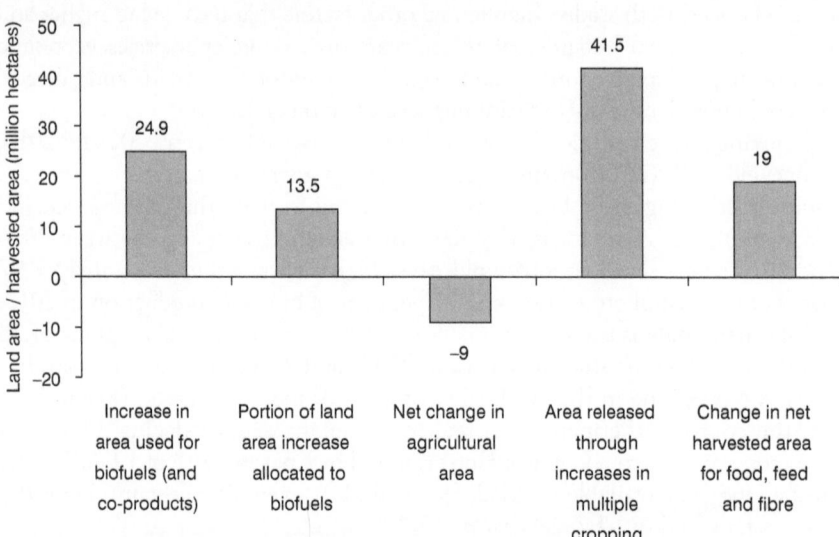

Figure 5.3 Changes in land use for biofuel crops and other agricultural crops across Brazil, USA, Indonesia, Malaysia, China, South Africa, Mozambique and the EU, 2000–2010

Source: Langeveld et al. (2014b)

Policy responses

Just as the issue of food security raises a diverse set of concerns and modelling studies have produced a diverse set of results, there is also diversity in the solutions proposed. Many of the options explored by policy-makers are similar to those canvassed in Chapters 2, 3 and 4 for dealing with the issues of climate change, deforestation and restoration. These include direct regulation to control what can be done with land or with the products produced from it, supply chain interventions that attempt to harness the power of the market to influence where energy crops are grown, and the strategic promotion of alternative feedstocks that are seen to pose less of a threat to food security. This section analyses each major policy option for protecting and enhancing food security, including examples of where they have been used or proposed. Lastly, the potential to move beyond the notion of "food vs fuel" is explored, with consideration given to integrated land use options that could simultaneously produce bioenergy while enhancing food security.

Regulating the use of land and agricultural commodities as feedstocks

Following the rapid rise in global food prices in 2007–8 and the concern that biofuel expansion was a key cause, many of the proposed solutions involved restrictions on the use of land and other resources for biofuel production. Prominent proposals at that time included a global moratorium on further biofuel expansion (Eide, 2008), the restriction of biofuel feedstocks to wastes only (Searchinger et al., 2008) and limiting any further expansion of biofuel feedstocks to "idle land" that was not currently being used to produce food or for any other productive purpose (Gallagher, 2008). However, each of these options present significant challenges, including the need for international cooperation to ensure that production does not simply move across national boundaries, the need for accounting systems to keep track of feedstocks that are readily interchangeable between the food and biofuel sectors, the need to define terms such as "idle land" and the risk that restrictive land use rules could preclude some forms of energy cropping that have potential benefits for ecosystem health, socio-economic wellbeing and even food security itself.

The influential Gallagher Review into the indirect effects of biofuels, commissioned by the UK government in 2008, concluded that the "optimum solution" to prevent food insecurity and deforestation from biofuel expansion was a global land use planning agreement. Such an agreement would direct biofuel feedstock production to areas where it is most appropriate, which for Gallagher was "idle land". The most obvious difficulty with such a plan, as acknowledged by Gallagher, would be reaching agreement among so many different nations with diverse views and interests regarding food security, economic development, farmer incomes, climate change, protection of biological diversity and other concerns.

In the absence of a global land use planning agreement, national or sub-national governments may choose to implement restrictions on land use to protect food production. There are a number of examples from around the world where restrictive zoning laws have been applied to protect agricultural land from conversion to non-agricultural uses. The US state of Oregon and the Canadian province of British Columbia have each implemented zoning rules aimed at slowing the loss of farming land to urban expansion. Similarly, the Australian state of New South Wales (NSW) has introduced a strategic regional land use planning process that is aimed at identifying prime agricultural areas at risk from mining or coal seam gas development. Such restrictions are generally not absolute, but may trigger assessment procedures. For example, the NSW arrangements require that any mining or coal seam gas activities within 2 km of strategic agricultural land undergo assessment by an independent panel (Department of Planning and Infrastructure, 2012).

While zoning rules such as those employed in Oregon, British Columbia or NSW could be used to restrict the expansion of non-edible energy crops like jatropha or willow, such an approach is more problematic for crops that can be used for either food or fuel. Unlike a change from farming to mining or urban expansion, a shift in land use from growing soy for food to growing soy for biodiesel is unlikely to result in any discernible change in land management practices. Furthermore, if the agricultural produce is sold into a global market, the change in end use from food to fuel could occur on the other side of the world without the land manager even being aware of it.

One option for dealing with the interchangeability issue is to apply zoning rules across entire feedstock categories, as Brazil has done by introducing zoning rules to control the future expansion of sugarcane and oil palm (Leopold, 2010). However, this system has been designed to limit negative environmental impacts like deforestation and is not capable of controlling the amounts of each crop that are used for food or fuel. It could be applicable for crops that are non-edible or rarely used for food (e.g. willow), but such crops are not usually the main source of concern in the food versus fuel debate. A second option would be for governments to introduce a product tracing system that keeps track of which biofuels have been produced from which feedstocks grown on which land. However, such a system would be extremely complex and costly and would require international coordination given the extensive global trade in biofuels and their feedstocks. A third option, which is the one that has been employed more widely to date, is to address the issue at the point of feedstock conversion by placing limits on which crops can be used for biofuel production.

China is the most prominent example of a national government which has moved to restrict the use of certain crops for biofuels, most notably corn, in response to concerns about food security. After actively promoting the use of corn for ethanol production over the period 1999–2008, the Chinese Government changed tack, responding to concerns around rising grain prices by halting the construction of new corn-to-ethanol plants (Zhong et al., 2010). In many ways, restrictions of this nature are simply an extension of the market interventions that are often employed around the world to protect domestic supplies of key staples, such as the export restrictions on wheat, rice and other crops introduced by China, India, Russia and a number of other countries following the 2007–8 food crisis.

The ability of the Chinese Government to intervene quickly to stop the expansion of grain-based ethanol across the country was enhanced by the fact that the four companies involved in ethanol production were all state-owned. Furthermore, biofuel development did not cease altogether, with the focus instead shifting to crops such as cassava and sweet sorghum (which are predominantly grown for animal feed rather than direct human consumption), as well as to advanced biofuels from cellulosic biomass and algae. Notably, the ban on new facilities does not directly restrict the amount of grain that can be used for ethanol, with the tonnages of corn and wheat used for ethanol in China both forecast to increase by over 30 per cent between 2010 and 2020 (Langeveld et al., 2014a). Thus, the Chinese experience offers a nuanced model for use in times of temporary food insecurity, whereby restrictions are aimed at slowing rather than halting biofuel growth and are combined with efforts to find alternative feedstocks.

While restrictions on the use of certain feedstocks for biofuels may be an effective way to halt or slow the diversion of crops away from food markets, care should be taken to ensure that all pressures on food production are considered equally. Other land use activities such as plantation forestry, cotton-growing, mining and urban expansion can also have direct impacts on the amount of land used for food production, as well as indirect impacts on food prices. Land use modelling undertaken by Langeveld et al. (2013) showed that, across the major biofuel-producing countries they reviewed, only in the US did biofuel expansion account for more than half of the net loss of land used for food, feed and fibre between 2000 and 2010. Biofuel expansion was not the major factor in the loss of agricultural land in the EU and China, where urbanisation, industrialisation and infrastructure development played key roles in the conversion of land.

Aside from the diversion of agricultural land to urbanisation and industrial development, increased production of cash crops for export can also have a negative effect on local food security. This impact is not restricted to biofuel crops and can occur even if the exports in question are food crops. A recent review of biofuels and food security by the International Development Institute found insufficient evidence to support the singling-out of biofuels for special treatment, stating that:

> Existing studies suggest that the impact of biofuels on food security may not differ markedly from that of other agro-industrial crops. Other factors may be more important than the crop itself in avoiding negative outcomes: the way that land is made available for projects; the project design and the models of production used; the use of existing safeguards and best practice in project design and land acquisition.
>
> (Locke and Henley, 2014, p. 1)

Chapter 6 looks at several of the key issues highlighted in the above quote, particularly the way that land is made available for projects and what model of production is used (e.g. broad-scale vs smallholder production). A lack of secure land tenure can be a key cause of local food insecurity, regardless of whether the new project in question involves energy cropping, the production of food crops for export, or a non-agricultural use such as mining, urban expansion or infrastructure development.

One of the problems with broad-based limits on biofuel production, either through restrictions on land use or on the use of certain feedstocks, is that they may pose a barrier to certain types of energy cropping that can actually deliver environmental and social benefits. For example, the Gallagher Review's proposal to restrict biofuel expansion to "idle land" would pose a major barrier for projects that seek to restore or protect degraded or vulnerable farmland, such as the Oil Mallee Project in Western Australia, discussed in Chapter 4. Figure 5.4 highlights this conflict by comparing the assumptions underpinning Gallagher's "idle land" proposal with the assumptions underpinning the development of an oil mallee industry in WA.

The conversion of small strips of a wheat field to mallee may result in a short-term decline in wheat production due to a loss of cropping area, but, if this helps to mitigate dryland salinity, the long-term result is likely to be higher levels of food production than if the area remained completely under wheat. Even the assumption that planting mallee would lead to a short-term reduction in food

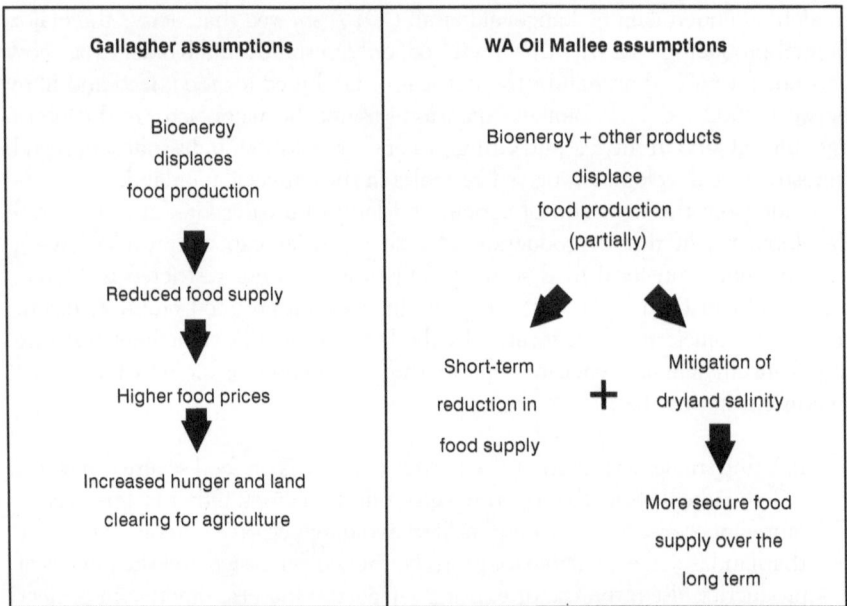

Figure 5.4 Comparison between the assumptions underpinning the approaches of Gallagher (2008) with those underpinning the West Australian Oil Mallee Project

supply may be overly pessimistic in some cases, as mallee belts can enhance wheat yields by reducing wind speeds at ground level (Abadi and Cooper, 2004).

Given the problems with dryland salinity in Western Australia, a blanket rule banning bioenergy plantations that displace food production could prove counter-productive in terms of food security, landscape health and even indirect land use change (as wheat production lost to salinity may be replaced by land clearing elsewhere in the world). In practice, any requirement for biofuel producers to use idle land would need to have a range of exemptions to cover cases such as oil mallee. The UK Government pursued the idle land concept for a brief period from 2008 to 2010, requiring biofuel producers to report on the type of land used to source their feedstocks. However, they subsequently adopted an approach consistent with the EU's Renewable Energy Directive, which requires biennial reports of the impact of EU biofuel targets on global food security, promotion of biofuels from wastes and cellulosic feedstocks and the use of certification schemes that comply with the EU sustainability criteria. These approaches are discussed below.

Supply chain approaches – biofuel certification

As with issues of deforestation, restoration and climate change, voluntary certification of biofuel producers offers a means of promoting biofuels that avoid negative impacts on food security. However, to do this, biofuel certification schemes have had to step away from the approach that has traditionally been used in agriculture and forestry standards, under which food security is not explicitly mentioned. For example, two of the prominent sustainability schemes looked at in this book so far, those of the Forest Stewardship Council (FSC) and the Sustainable Agriculture Network (SAN), do not contain any criteria specifically related to food security. The closest they come are general obligations to identify and consider the interests of local populations (SAN) and to maintain or enhance social wellbeing and community needs (FSC). This is despite the fact that a wide range of cash crops, including food crops grown for export, have the potential to exacerbate local food insecurity if an inappropriate approach is taken to land acquisition and production (Locke and Henley, 2014).

In contrast to the FSC and SAN standards, those of the Roundtable on Sustainable Biomaterials (RSB) have one whole criterion dedicated to food security. This reflects the greater emphasis that has been placed on food security in the biofuel sector due to the food vs fuel debate. Under the RSB standards, biofuel producers must assess risks to local food security from their operations and mitigate any negative impacts. Furthermore, if operating in a food insecure region, they must actively enhance the food security of directly affected stakeholders through measures such as setting aside land for food production or increasing yields. This obligation goes beyond simply *maintaining* local food security to actively *enhancing* it. Notably, the RSB does not require biofuel producers to consider impacts on global food security, arguing that such impacts are "indirect", "macro-level" and "beyond the control of the individual farmer or biofuels producer seeking certification" and thus are best dealt with by engaging with intergovernmental agencies and other stakeholders rather than through voluntary certification schemes (RSB, 2010, p. 4).

The RSB standards are one of the standards endorsed by the EU as complying with its sustainability criteria under the Renewable Energy Directive. This means that biofuel producers certified by the RSB are eligible to have their biofuels counted under the EU's biofuel targets. However, it is important to note that the EU criteria do not require biofuel producers to enhance food security or mitigate any negative impacts on food security. These are additional requirements that the RSB has chosen to include in their standards. Partly this reflects a reluctance on the part of the EU to include social criteria in their sustainability requirements, which would be difficult to implement under World Trade Organization (WTO) rules (Charnovitz et al., 2008). However, it is also reflects the fact that the RSB has chosen to place an emphasis on food security that other certifying bodies have not.

German and Schoneveld (2011) reviewed the RSB standards alongside six others endorsed under the EU RED, finding that the RSB had the most substantial protections relating to food security. Of the seven schemes reviewed, four included no coverage of food security issues at all – those produced by the Round Table for Responsible Soy, Bonsucro (focused on sugarcane), Abengoa (a Spanish ethanol producer) and the Biomass Biofuels Sustainability Voluntary Scheme (a consortium of French biofuel companies). Two of these standards are biofuel-specific and the other two cover soy and sugarcane produced for a variety of uses. The RSB's decision to make food security a central issue in their standards undoubtedly reflects the prominence of the food vs fuel issue in recent years, but it should not be taken as proof that biofuel feedstocks automatically represent a greater threat to food security than other crops. Soy and sugarcane production can also have an impact on local food security, as can the production of the feedstocks covered by the French and Spanish biofuel schemes. The emphasis placed on different issues under different standards inevitably reflects the values of the organisations that produce them.

A 2013 report to the FAO (Elbehri et al., 2013) highlights another key issue around biofuel certification schemes – that they are generally not structured to be inclusive of small-scale producers. The fact that certification schemes such as that of the RSB require costly, complex and intensive information systems and management capacities makes them "largely out of reach for small-scale producers" (Elbehri et al., 2013, p. xvi). As such, the RSB standards may help to ensure that large producers moving into an area consider their impact on local food security, but are less useful for small producers seeking to combine food production with energy cropping by accessing markets such as the EU. Biofuel policy design approaches that specifically aim to include small-scale producers are considered in Chapter 6 on land rights and community impacts.

Promotion of alternative feedstocks and multifunctional land use options

Concerns about food security, along with climate change and deforestation, have led to calls for future bioenergy production to be sourced from wastes or residues of other processes, or from cellulosic crops such as willow, eucalypts or perennial grasses. As discussed in Chapter 2, the EU promotes the use of such feedstocks by allowing biofuels from these sources to be "double-counted" against targets, meaning that fuel suppliers can fulfil their biofuel obligations more easily if they use these fuels. The US Government has also been increasing support for advanced biofuels produced from wastes and cellulosic feedstocks by progressively increasing the requirement for fuel suppliers to use such fuels. However, food security has not been cited as a key goal underlying the US incentives and, at any rate, the US EPA has been forced to lower its requirements for advanced biofuels on an annual basis due to a lack of supply (Environmental Protection Agency, 2013).

As discussed in Chapter 2, the EU has been debating a proposal since 2012 that would increase the incentive to use feedstocks from wastes and cellulosic crops. This proposal makes it clear that food security is one of the key concerns in moving away from first-generation biofuels, alongside concerns around greenhouse gas emissions and deforestation from indirect land use change (European Commission, 2012). If the proposed amendments are adopted, first-generation biofuels would be limited to the share of the biofuel market they occupied in 2011 and would no longer be eligible for subsidies after 2020. Cellulosic crops, which are seen to produce less competition by being suitable for lower-quality land would continue to be promoted by double-counting them against RED targets. Waste feedstocks, which do not require new land to be used at all, would be given additional support through a quadruple-counting approach.

While cellulosic energy crops may be less likely to compete with food crops, it is possible to go one step further and promote crops that actually enhance food security at the same time as producing biofuel feedstocks. At present, there are no clear examples of policy measures designed specifically to promote biofuel feedstocks that actually enhance food security. As with the delivery of ecological restoration outcomes discussed in Chapter 4, one key barrier to multifunctional cropping systems for energy, food and ecosystem enhancement is the lack of commercially viable models at present. However, a range of options cited in Chapter 4 are at various stages of development. Poplar and willow in Europe have been used to rehabilitate land contaminated with heavy metals. Oil mallee in Western Australia has the potential to enhance long-term food production from wheat fields while replacing fossil fuels and combating dryland salinity. Jatropha has the potential to complement grazing or cropping systems on marginal land by offering windbreaks, soil protection and the sustainable harvest of oil-rich seeds for biodiesel production. Even well-established crops like sugarcane and oil palm can offer benefits over pre-existing land uses if established with care, as shown by the examples from Florida and Brazil cited in Chapter 4.

Figure 5.5 Multifunctional agroforestry systems involving the intercropping of eucalypts with cattle (left) and soy (right) in Brazil
Source: Couto et al. (2011)

Berndes et al. (2011) argues that there are great untapped opportunities to increase productivity from bioenergy crops through the trialling of new species, enhanced breeding methods and multifunctional production systems. For example, much work has been undertaken in Brazil on the integration of eucalyptus plantations for energy alongside cattle grazing or soybean cropping (Figure 5.5). Similarly, in Australia there has been a substantial amount of research and development into new agroforestry systems that combine food and bioenergy production, such as the use of golden wreath wattle (*Acacia saligna*) as a phase crop to enhance agricultural yields while producing biomass for energy (Sudmeyer et al., 2013).

While the lack of commercially proven examples of multifunctional bioenergy systems has meant there has not been the impetus for multifunctional policy incentives to date, it is also the case that the introduction of such incentives could spur the commercialisation of some of the land use options discussed above. If the use of waste feedstocks in the EU can be incentivised through a quadruple-counting model, it may also be possible to implement a similar arrangement to promote perennial cropping systems that actually enhance food security. Similar incentives could be provided for electricity generation from perennial energy crops using policy options such as mandates and feed-in tariffs (discussed in more detail in Chapter 8).

Conclusion

Food security is a complex issue and vigilance is required on the part of bioenergy producers and policy-makers to ensure that increased bioenergy production does not exacerbate the problem of food insecurity. This may require targeted interventions at times of high food prices or food insecurity, such as a slowdown in biofuel targets or a shift away from first-generation feedstocks. These measures will also need to be complemented by measures that are not specific to biofuels, such as ensuring that vulnerable people have access to sufficient food through secure land tenure, adequate incomes and food aid at critical times. However, there are also promising signs that food and fuel production can be successfully integrated, from land use modelling showing that sufficient land is available for both food and fuel (Dornburg et al., 2010), to the role that multiple cropping can play in increasing productivity (Langeveld et al., 2013), to the practical examples highlighted in this chapter of food and fuel being combined in multifunctional production systems.

At present, the lack of incentives for multifunctional bioenergy systems represents a significant missed opportunity for promoting approaches that simultaneously provide food and fuel (as well as other co-benefits for biodiversity, soils and water quality). However, as the use of cellulosic feedstocks continues to grow through the application of new technologies and the incentives on offer, this opportunity will hopefully be capitalised on by governments and non-government actors such as the RSB. While a cautious approach is required for first-generation biofuels to prevent the rapid growth in feedstock demand from swamping food markets, targeted incentives could be progressively introduced to promote forms of bioenergy that go further than simply being "not unsustainable" and actively promote sustainability on a number of levels simultaneously.

As familiarity with emerging bioenergy crops increases and policy-makers devise ways of incentivising multifunctional land uses, it may be possible to move beyond the notion of food vs fuel that has dominated this issue to date. Instead, we may begin to aim for energy cropping systems that do not simply avoid competition with food, but actively seek to enhance multiple outcomes over the long-term. The policy options required to achieve this are discussed in more detail in Chapters 8 and 9, where an integrated approach is taken that considers food security alongside climate change mitigation, ecological restoration and the other potential impacts of energy cropping. The following chapter on land rights and community impacts also considers ways that energy cropping systems can be better designed to benefit smallholders and achieve local community objectives, including local food security.

References

Abadi, A. and Cooper, D. (2004) "A case study of the economics of alley farming with oil mallees in Western Australia using the *Imagine* framework", in Graham, T. W., Pannell, D. J. and White, B. (eds), *Dryland Salinity: Economic Issues at Farm, Catchment and Policy Levels*, Cooperative Research Centre for Plant-Based Management of Dryland Salinity, Perth, 103–18.

AETS (2013) *Assessing the Impact of Biofuels Production on Developing Countries from the Point of View of Policy Coherence for Development*, Framework Contract Commission, European Union, Brussels.

Ajanovic, A. (2011) "Biofuels versus food production: Does biofuels production increase food prices?", *Energy*, 36: 2070–76.

Anseeuw, W., Wily, L. A., Cotula, L. and Taylor, M. (2012) *Land Rights and the Rush for Land: Findings of the Global Commercial Pressures on Land Research Project*, International Land Coalition, Rome.

Bauen, A., Berndes, G., Junginger, M., Londo, M., Vuille, F., Ball, R., Bole, T., Chudziak, C., Faaij, A. and Mozaffarian, H. (2009) *Bioenergy: A Sustainable and Reliable Energy Source*, IEA Bioenergy Secretariat, Rotorua.

Berndes, G., Bird, N. and Cowie, A. (2011) *Bioenergy, Land Use Change and Climate Change Mitigation*, IEA Bioenergy, Rotorua.

Brown, L. R. (2008) *Plan B 3.0: Mobilizing to Save Civilization* , W. W. Norton and Company, New York.

Charnovitz, S., Earley, J. and Howse, R. (2008) *An Examination of Social Standards in Biofuels Sustainability Criteria*, International Food and Agricultural Trade Policy Council, Washington, DC.

Couto, L., Nicholas, I. and Wright, L. (2011) *Short Rotation Eucalypt Plantations for Energy in Brazil*, IEA Bioenergy Secretariat, Rotorua.

Dehue, B., Meyer, S. and Hamelinck, C. (2007) *Towards A Harmonised Sustainable Biomass Certification Scheme*, Ecofys Netherlands BV, Utrecht.

Dornburg, V., Faaij, A., Verweij, P., Langeveld, H., Ven, G. v. d., Wester, F., Keulen, H. v., Diepen, K. v., Meeusen, M., Banse, M., Ros, J., Vuuren, D. v., Born, G. J. v. d., Oorschot, M. v., Smout, F., Vliet, J. v., Aiking, H., Londo, M., Mozaffarian, H. and Smekens, K. (2008) *Biomass Assessment: Assessment of Global Biomass Potentials and Their Links to Food, Water, Biodiversity, Energy Demand and Economy*, Netherlands Research Programme on Scientific Assessment and Policy Analysis for Climate Change (WAB), Bilthoven.

Dornburg, V., van Vuuren, D., van de Ven, G., Langeveld, H., Meeusen, M., Banse, M., van Oorschot, M., Ros, J., Jan van den Born, G., Aiking, H., Londo, M., Mozaffarian, H., Verweij, P., Lysen, E. and Faaij, A. (2010) "Bioenergy revisited: key factors in global potentials of bioenergy", *Energy and Environmental Science*, 3: 258–67.

Eide, A. (2008) *The Right to Food and the Impact of Biofuels (Agrofuels): Advanced Copy*, Food and Agriculture Organization of the United Nations, Rome, 54p.

Elbehri, A., Segerstedt, A. and Liu, P. (2013) *Biofuels and the Sustainability Challenge: A Global Assessment of Sustainability Issues, Trends and Policies for Biofuels and Related Feedstocks*, Food and Agriculture Organization of the United Nations (FAO), 174p.

Energy Information Administration (2015) "International energy statistics: biofuels production", www.eia.gov/cfapps/ipdbproject/IEDIndex3.cfm?tid=79&pid=79&aid=1 (accessed 9 March 2015).

Environmental Protection Agency (2013) *EPA Proposes 2014 Renewable Fuel Standards, 2015 Biomass-Based Diesel Volume*, United States Environmental Protection Agency, Washington, DC.

European Commission (2012) *Proposal for a Directive of the European Parliament and of the Council Amending Directive 98/70/EC Relating to the Quality of Petrol and Diesel Fuels and Amending Directive 2009/28/EC on the Promotion of the Use of Energy from Renewable Sources*, European Commission, Brussels.

Fischer, G., Hizsnyik, E., Prieler, S., Shah, M. and Velthuizen, H. v. (2009) *Biofuels and Food Security: Implications of an Accelerated Biofuels Production*, OPEC Fund for International Development (OFID), Vienna.

Gallagher, E. (2008) *The Gallagher Review of the Indirect Effects of Biofuels Production*, Renewable Fuels Agency, London.

German, L. and Schoneveld, G. (2011) *Social Sustainability of EU-Approved Voluntary Schemes for Biofuels: Implications for Rural Livelihoods*, Center for International Forestry Research (CIFOR), Bogor.

Hoogwijk, M., Faaij, A., de Vries, B. and Turkenburg, W. (2009) "Exploration of regional and global cost-supply curves of biomass energy from short-rotation crops at abandoned cropland and rest land under four IPCC SRES land-use scenarios", *Biomass and Bioenergy*, 33: 26–43.

Kim, S. and Dale, B. E. (2011) "Indirect land use change for biofuels: testing predictions and improving analytical methodologies", *Biomass and Bioenergy*, 35: 3235–40.

Langeveld, J. W. A., Dixon, J., Keulen, H. v. and Quist-Wessel, P. M. F. (2013) *Analysing the Effect of Biofuel Expansion on Land Use in Major Producing Countries: Evidence of Increased Multiple Cropping*, Biomass Research, Wageningen.

Langeveld, J. W. A., Dixon, J. and van Keulen, H. (2014a) "Biofuel production in China", in Langeveld, J. W. A., Dixon, J. and Keulen, H. v. (eds), *Biofuel Cropping Systems: Carbon, Land and Food*, Routledge, Abingdon, 195–207.

Langeveld, J. W. A., Dixon, J. and van Keulen, H. (2014b) "Impact on land and biomass availability", *In:* Langeveld, J. W. A., Dixon, J. and Keulen, H. v. (eds.) *Biofuel Cropping Systems: Carbon, Land and Food*, Routledge, Abingdon, 239–250.

Leopold, A. (2010) *Agroecological Zoning in Brazil Incentivizes More Sustainable Agricultural Practices*, The Economics of Ecosystems and Biodiversity (TEEB), Geneva.

Locke, A. and Henley, G. (2014) *Biofuels and Local Food Security: What Does the Evidence Say?*, Overseas Development Institute, London.

Mitchell, D. (2008) *A Note on Rising Food Prices*, World Bank Development Prospects Group, Washington, DC.

Monbiot, G. (2007) "An agricultural crime against humanity", www.monbiot.com/2007/11/06/an-agricultural-crime-against-humanity.

Oxfam International (2007) *Bio-fuelling Poverty: Why the EU Renewable-Fuel Target May Be Disastrous for Poor People*, Oxfam, Oxford.

Rosillo-Calle, F., Groot, P. D., Hemstock, S. L. and Woods, J. (2015) *The Biomass Assessment Handbook: Energy for a Sustainable Environment* (2nd ed.), Routledge, London.

RSB (2010) *Global Principles and Criteria for Sustainable Biofuels Production: Version 2.1*, Ecole Polytechnique Federale de Lausanne, Lausanne.

Schoneveld, G. C. (2010) *Potential Land Use Competition from First-Generation Biofuel Expansion in Developing Countries*, Center for International Forestry Research, Bogor.

Searchinger, T., Heimlich, R. A., Houghton, F., Dong, A., Elobeid, J., Fabiosa, S., Tokgoz, D. Hayes and Yu, T. (2008) "Use of US croplands for biofuels increased greenhouse gases through land-use change", *Science*, 319: 1238–40.

Sudmeyer, R. A., Mazanec, R., Rogers, D., Simons, J. and Daniels, T. (2013) *Integrating Food and Energy Systems*, Rural Industries Research and Development Corporation, Canberra.

UN Energy (2007) *Sustainable Bioenergy: A Framework for Decision Makers*, United Nations, New York.

USDA (2010) *USDA Agricultural Projections to 2019*, United States Department of Agriculture, Washington, DC.

USDA (2015) "Downloadable datasets: grains", http://apps.fas.usda.gov/psdonline/psdDownload.aspx (accessed 4 March 2015).

Zhong, C., Cao, Y.-X., Li, B.-Z. and Yuan, Y.-J. (2010) "Biofuels in China: past, present and future", *Biofuels, Bioproducts and Biorefining*, 4: 326–42.

Zilberman, D., Hochman, G., Rajagopal, D., Sexton, S. and Timilsina, G. (2013) "The impact of biofuels on commodity food prices: assessment of findings", *American Journal of Agricultural Economics*, 95: 275–81.

Chapter 6

Land rights and community impacts

The expansion of energy cropping across the globe has the potential to produce a variety of impacts for local communities, both positive and negative. Some of these impacts have already been touched on in Chapter 5, which dealt with the issue of food security, and Chapters 3 and 4, which dealt with changes to the ecosystem services on which local communities may depend. This chapter takes the discussion of social impacts further, dealing first with land tenure and other resource use rights, before moving on to the other socio-economic impacts that can be felt by local communities as energy cropping expands. These other impacts, including job creation, workers' rights, shifts in rural population and changes to the aesthetic values of areas affected by energy cropping, often play out very differently in developing countries than they do in developed ones. The chapter concludes with a discussion of how policy measures may be designed to guard against negative social impacts while enhancing the potential benefits for farmers and local communities.

Land tenure and resource use rights

As with issues of deforestation and food security, the impact that energy cropping may have on security of land tenure can be either direct or indirect. An example of a direct impact would be the displacement of local people by a company seeking to use their land for the production of an energy crop, such as oil palm for biodiesel or sugarcane for ethanol. An example of an indirect impact would be the growing demand for biofuels driving up prices for agricultural products and therefore creating an incentive for investors to develop new cropping areas by pushing vulnerable people off their land. In this latter case, the land lost by local people may actually end up being used for the production of food or animal feed, rather than energy crops. However, the growth in biofuel demand will have played a role in setting off the "domino effect" that led to this land use change. The focus of this chapter is primarily on the direct role that energy cropping might play in the displacement of vulnerable people, but it is important to keep in mind this potential for indirect impacts when considering which of the proposed solutions are likely to be most effective.

Concerns around the displacement of local people for energy cropping are most prominent in developing countries. This is due to combination of factors, including a concentration of poverty among many rural and indigenous populations, a reliance on customary or traditional approaches to land tenure that can leave local people without legal protections, difficulties with law enforcement and corruption, and significant power differentials between local landholders and the business or political elites who often promote and benefit from plantation development. However, the situation varies from country to country and the displacement of local people is far from being an inevitable consequence of new energy cropping activities in developing countries.

A 2012 report from the International Land Coalition (ILC), a global alliance of intergovernmental and non-government organisations, highlighted that the peaking of global food prices in 2007–8 sparked something of a "land rush" in many developing countries (Anseeuw et al., 2012). In particular, the ILC's concerns relate to large-scale land acquisitions, which have emerged as the dominant vehicle for investing in agriculture in many developing countries rather than investment in smallholder production. The report highlighted that Africa had emerged as the primary target of this land rush, accounting for around half of all verified land transactions, followed by Asia.

The rural poor bear disproportionate costs from large-scale land transactions, with women particularly vulnerable. A lack of legal recognition for customary land tenure, along with corrupt and unaccountable decision-making and a lack of policy support for smallholder agriculture can compound the risks posed by large-scale land acquisition. Lands under communal or customary tenure are often viewed by political and business elites as under-utilised, idle or marginal, in contrast to local people who may value such areas for fuelwood collection, hunting, gathering of food and medicines, animal grazing (especially during droughts), shifting cultivation and/or spiritual purposes. As discussed in Chapter 5, the loss of access to these lands can exacerbate food insecurity at the local scale.

The ILC is a key partner in the "Land Matrix", a database of large-scale land transactions that have occurred since 2000, compiled from a range of mostly unofficial sources. While transactions can be hard to verify and the intended purpose for the land is not always known, the ILC's 2012 report (Anseeuw et al., 2012) cited evidence from the Land Matrix that biofuel demand represents a major factor in such transactions. Across all of the land transactions investigated for the report that were able to be verified and categorised according to their intended purpose, more than half of the total land area was reported to be intended for the production of biofuel feedstocks.

Apart from the rising demand for biofuels, there are other factors that can make biofuel feedstocks particularly suited to large-scale land transactions. The need for economies of scale, the need for investors in distant markets to be able to monitor and manage plantations from afar, and the need to integrate feedstock production with processing sites and export infrastructure are all factors that have been argued to make biofuel feedstocks particularly suited to large-scale monocultures (Eide, 2008). While these economic pressures are by no means unique to biofuels, and it is true that other models of biofuel production may be feasible, it is important to recognise that such pressures do exist and can work against attempts to foster smallholder production that is sensitive to the needs of local communities and ecosystems.

Oil palm expansion in southeast Asia has been highlighted as a concern for many smallholders and indigenous people with insecure land tenure. Indonesia's constitution grants the state considerable powers to take control of land for development projects that are deemed to be in the national interest, with poor protection of customary land rights and a lack of recognition of the need for free, prior and informed consent by local communities. The use of this power has led to considerable conflict around the development of land, with investors often relying on political connections to open up new areas (Sheil et al., 2009; Colchester, 2011).

In recent years, Indonesia has moved towards an approach based on joint ventures that require the consent of local landholders, following a strategy employed in Sarawak, Malaysia. However, concerns have arisen around this approach, with a lack of contractual security for landholders and uncertainty about the areas of land given up and the benefits to which landholders are entitled (Colchester, 2011). This lack of certainty clearly works against local communities, but it can also work against oil palm developers and the environment. Where land rights are contested, oil palm developers may have a perverse incentive to target lightly-populated forested areas over previously cleared agricultural areas in order to reduce the number of landholders they must negotiate with (Sheil et al., 2009).

Sugarcane for ethanol and jatropha for biodiesel are other energy crops that have been implicated in large-scale land acquisitions with detrimental effects on local communities. In the Philippines, the development of sugarcane ethanol has been driven by a combination of Chinese investment and a government target of developing 2 million hectares of land considered idle or underutilised (Ravanera and Gorra, 2011). Jatropha has also been a target crop for the Philippines Government, with Ravanera and Gorra citing concerns that some of these deals have involved very small payments to local landholders who risk the long-term security of their land tenure by handing over their land for ten years or more to plantation developers.

Despite being yet to achieve widespread success as a biofuel feedstock, jatropha has been commonly cited as a driver of land dispossession, particularly in Africa. Friends of the Earth (2010) cite examples from Tanzania, Mozambique, Ghana and Zambia where local communities claim to have been cheated in deals with biofuel investors or have had their land taken without compensation. Biofuel production has also been at the centre of land deals in Madagascar, where large-scale land transactions have been highly controversial since the collapse in 2009 of a deal that would have seen some 1.3 million hectares of farmland (between 15% and 37% of all arable land in the country) transferred to the control of a Korean company. While that deal was brought down by community opposition (bringing down the Madagascan President in the process), an analysis by Ratsialonana et al. (2011) found that many land transactions were still underway or planned, with the majority being for biofuels produced from sugarcane, palm oil and particularly jatropha.

Cellulosic energy crops, such as trees and grasses for bioelectricity, have attracted less attention than biofuel crops with regards to land grabs and negative community impacts. However, certain projects have attracted controversy, such as the eucalypt plantations established by the Plantar Group in the Brazilian state of Minas Gerais for charcoal production. Part of this controversy relates to whether the land should be considered eligible for reforestation under the Clean Development Mechanism (CDM) of the Kyoto Protocol, but issues of land grabs and social exclusion have also been raised. The NGO Carbon Market Watch claims that the project has involved illegal dispossession of local people, exclusion of wood-collection activities from the site and worker exploitation. While the Plantar Group argue that the project has in fact delivered numerous socio-economic benefits for local people, they acknowledge that there have been some "adverse public reactions to this activity" (Carbon Market Watch, 2010, p. 5).

While the above examples highlight that bioenergy demand can increase risks of land dispossession for vulnerable landholders, it is important to remember that the problem is not unique to bioenergy. The displacement of local smallholders to make way for cash crops destined for foreign markets has long been an issue for a range of crops, including coffee, cocoa, bananas, sugar and plantation timber. Indeed, the aforementioned land deal that brought down the President of Madagascar was in fact for food crops rather than biofuels. Furthermore, a close look at the ILC's Land Matrix as it appears in March 2015 (Figure 6.1) indicates that biofuel feedstocks may no longer be as dominant in land deals as was reported in the 2012 ILC report (Anseeuw et al., 2012).

Of a total of 1930 transactions listed in the Land Matrix as of 12 March 2015, 22 per cent had biofuels listed as an intended purpose, compared with 41 per cent that had food crops listed as an intended purpose. Furthermore, out of a total of 53 million ha listed in the contract size column of the Land Matrix, biofuels were a listed intention for only 10 million ha. This is less than the 20 million hectares intended for food crops and the 14 million ha intended for wood and fibre products. It is possible that biofuels may be produced from some of the contracted land listed as "unspecified" or "non-food agricultural", but it is also likely that some of this land would be used for food crops, livestock grazing or wood and fibre production. Overall, the picture presented by Figure 6.1 suggests biofuel production is one of several drivers of large-scale land transactions worldwide, rather than being the dominant cause.

The argument that land rights issues are not unique to bioenergy crops is further supported by a comparison between prominent sustainability standards from the biofuels, agriculture and forestry sectors. Table 6.1 compares criteria relating to land rights and other social issues across the same three standards that were analysed in Chapters 3 and 5 – those of the Roundtable on Sustainable Biomaterials (RSB), the Forest Stewardship Council (FSC) and the Sustainable Agriculture Network (SAN). As with the analysis of food security in Chapter 5, the EU's Renewable Energy Directive has been left out due to the fact that the RED does not have prescriptive sustainability criteria covering land rights, workers' rights or other social factors.

As can be seen in Table 6.1, the RSB, FSC and SAN standards all have criteria relating to land rights, indicating that this is not an issue restricted to the biofuel sector. The RSB requirement to respect land rights has two sub-criteria, with biofuel producers first required to assess, document and establish the formal and informal rights to the land, before ensuring that any negotiations for compensation, acquisition or voluntary relinquishment of rights are underpinned by consent that is free, prior and informed. The FSC and SAN requirements are similar, although the FSC only requires consent to be "free and informed" and the SAN does not provide any specific details about the nature of the consent required.

Sustainability standards such as those of the RSB, FSC and SAN represent an alternative way of protecting land rights where government regulations are inadequate or difficult to enforce. The ability of national governments to adequately protect land rights is often compromised by their role as active facilitators of many, if not most, large-scale land transactions (Zoomers, 2010). This can be seen in some of the examples discussed previously, such as the Madagascan deal to transfer 1.3 million hectares of farmland to a Korean company or the powers granted by the Indonesian constitution to seize land for

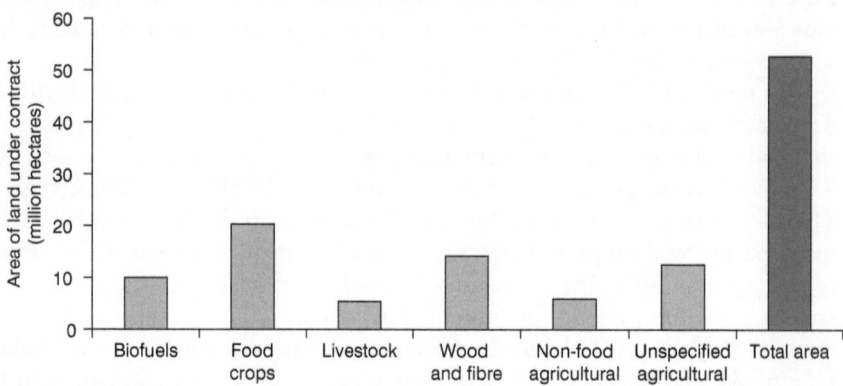

Figure 6.1 Intended purposes for land transactions listed under the ILC's Land Matrix as of March 2015

Source: Land Matrix (2015)

Note: The "Total area" is less than the sum of all other columns, as land transactions may have more than one intended purpose. Only land transactions with an area listed in the "contract size" column have been included in the analysis. Non-agricultural land uses (e.g. tourism, industry, conservation, carbon sequestration) are not shown in the graph.

Table 6.1 Sustainability criteria relating to land rights and other social impacts in the RSB, FSC and SAN standards

RSB 2010	FSC 2012	SAN 2010
Biofuel operations shall: • Not violate human rights or labour rights, and shall promote decent work and the well-being of workers • In regions of poverty, contribute to the social and economic development of local, rural and indigenous people and communities • Respect the existing water rights of local and indigenous communities • Respect land rights and land use rights	The organisation shall: • demonstrate that the legal status of the management unit, including tenure and use rights, and its boundaries, are clearly defined • identify and uphold indigenous peoples' legal and customary rights of ownership, use and management of land, territories and resources affected by management activities • maintain or enhance the social and economic well-being of workers • contribute to maintaining or enhancing the social and economic wellbeing of local communities	• The farm must not discriminate in its labour and hiring policies and procedures • Workers must receive legal remuneration greater than or equal to the regional average or the legally established minimum wage (whichever is greater) • It is prohibited to directly or indirectly employ workers under the age of 15 (apart from some exceptions for family farms) • Any type of forced labour is prohibited • Workers must have the right to freely organise and voluntarily negotiate their working conditions in a collective manner • The farm must prioritise the hiring of local people and local goods and services • The farm must collaborate with the development of the local economy and contribute to community costs • The farm must have a legitimate right to land use and tenure

Note: Social criteria relating specifically to food security have been excluded, as they were discussed in Chapter 5.

development projects in the "national interest". Sustainability standards and their associated certification schemes attempt to overcome this by harnessing the concerns of consumers, traders and processors, particularly those in developed countries, to ensure that biomass produced without respect for the rights of local people cannot enter the supply chain.

While sustainability standards seek to bypass ineffectual government regulation, a question remains about how effective they are themselves at preventing "land grabs". Analysis by Fortin and Richardson (2013) found that the RSB's land rights criteria had some significant strengths, such as the focus on all "land users" rather than just those with official tenure. This finding is supported by earlier analysis by German and Schoneveld (2011), which showed that the RSB standards were stronger on a range of social factors than several other biofuel standards approved by the EU. However, Fortin and Richardson also argue that the effectiveness of certification schemes in general is limited by an inability to audit all parts of a participant's supply chain and an inherent conflict between the need to discipline participants who fail to meet certain criteria and the need to "sell" the scheme to new participants (who may have a choice of other schemes with less stringent criteria).

While certification schemes may be unable to strictly enforce their criteria, Fortin and Richardson argue that the primary value of such schemes may lie less in their ability to enforce strict rules and more in their potential to enable scrutiny and increase corporate accountability in transnational commodity chains. This has three key elements in relation to land rights concerns. First, certification schemes may be able to bring to light many details of large-scale land transactions that would otherwise remain hidden from public view. Second, they help to address the asymmetries of power that often make it hard for local groups to challenge large multinational companies (they do this by including such companies as members of their "roundtable" structures and exerting pressure on them via their industry peers). Third, sustainability certification schemes can act as a testing ground for new rules and mechanisms and can put pressure on governments to change their regulations. For example, the RSB has been active in leading discussions around the promotion of biofuels from wastes and cellulosic feedstocks in order to reduce pressures on land availability. This has in turn contributed to the development of the proposed reforms to the EU RED that would encourage greater use of wastes and cellulosic feedstocks (European Commission, 2012).

Other social impacts of energy cropping

Apart from demonstrating that it is not just the biofuel sector that needs to be concerned with issues of land tenure, Table 6.1 also highlights that land tenure is only one of the social issues that can arise from the introduction of new cropping systems. Labour rights and socio-economic development also feature prominently in the standards of the RSB, FSC and SAN. On the one hand, these criteria highlight the potential for energy cropping, forestry and agriculture to lead to positive outcomes such as increased employment and new livelihood options. However, they also highlight the potential for negative impacts, such as the risk of worker exploitation and unequal distribution of impacts. Production models that engage smallholder farmers are likely to have the greatest socio-economic benefits, but can be more demanding in terms of coordination and management (Hilhorst, 2014).

In terms of labour rights, the SAN standards are more stringent than the FSC or RSB standards, with sub-criteria on factors such as freedom of association, discrimination, child labour, slave labour, acceptable wages, health and safety. The RSB and FSC standards cover some of these factors under broader terms such as the "well-being" of workers, but do not go into the same level of detail as the SAN standards. In part, this reflects the different origins of each set of standards and the priorities of the stakeholders involved in developing them. The SAN standards operate under the banner of the Rainforest Alliance, which has traditionally had a strong focus on "fair trade" and has targeted agricultural sectors such as coffee and cocoa, where labour rights issues have attracted international attention. In comparison, the dominant issues driving consumers to look for FSC-certified products have tended to be tropical deforestation and unsustainable forest harvesting.

The RSB has attempted to strike a balance between environmental, social and economic issues in its standards and has no doubt benefitted from having earlier standards such as those of the FSC and SAN to draw on. Some biofuel feedstock sectors have attracted considerable attention over labour rights issues, especially sugarcane production. However, it is rare to see labour rights discussed as a "biofuel issue" in the same way that issues like deforestation, competition with food production and land grabs often are.

In terms of economic development, the SAN standards appear to set the highest standard, requiring producers to "collaborate with the development of the local economy". This implies that it is not enough to simply maintain conditions, which is the benchmark that is often used in relation to environmental criteria such as soil health or water quality. The RSB standards have similar requirements to the SAN, but they only apply in "regions of poverty", where biofuel operations must improve the socioeconomic status of local stakeholders impacted by biofuel operations and provide special measures for women, youth and indigenous people. The FSC sets a lower bar overall, with operators required to "maintain or enhance" the well-being of workers and communities (i.e. not make things worse). However, the FSC standard does include a requirement that operators must actively promote local employment and processing of forestry products.

It is arguable that standards such as those of the RSB have a limited role to play in maximising the socio-economic benefits of energy cropping, such as job creation and enhanced livelihood opportunities for farmers and local service providers. This is because such standards are generally developed in response to perceived threats, such as deforestation, greenhouse gas emissions, land-grabbing, food insecurity or worker exploitation. As such, the primary motivation for most biofuel producers to adopt these standards is to gain recognition for not contributing to threats (rather than to be seen as a driver of positive change). Even in the case of the EU RED, where energy cropping is actively promoted, the primary motivation relating to positive change is to mitigate climate change rather than to maximise socio-economic benefits. Furthermore, the costs and information requirements to achieve certification can pose a barrier to small-scale producers, who are arguably the most important target group when it comes to promoting socio-economic benefits (Elbehri et al., 2013).

The emergence of the "fair trade" movement in recent years demonstrates that for some products there are consumer segments willing to make purchasing decisions based on a product's contribution to enhancing social outcomes for local communities. However, fair trade motivations tend to be most prominent around food products, especially those that have been subject to negative press around "unfair trade" (e.g. worker exploitation or low prices paid to farmers). Notable examples include coffee, cocoa and bananas, which are common products certified by the Rainforest Alliance using the SAN standards.

Food products make up 14 of the 22 product categories certified by the UK-based Fairtrade Foundation, with most non-food categories being other products that consumers have an intimate long-term connection to, such as beauty products, cotton clothing and gold jewellery (Fairtrade Foundation, 2015). In comparison, biofuels are not directly consumed by people (and are often "hidden" in blended fuel), are a relatively new fuel for most consumers (who usually have alternatives readily available) and are more commonly associated with issues such as deforestation, greenhouse gas emissions and food insecurity rather than unfair trade. As such, it is questionable whether many consumers would actively seek out certified biofuels out of a desire to contribute to positive social development in the same way they might for other fair trade products.

The fact that the RSB has chosen to include criteria around socio-economic enhancement in their standards demonstrates a desire to promote a holistic notion of sustainability. However, it is arguable that to some extent these measures are "piggy-backing" on the criteria relating to lifecycle greenhouse gas savings, forest protection and local food security, which are more likely to be the main motivations for a biofuel producer to seek RSB certification.

Social factors affecting energy cropping in developed countries

While developing countries are often the focus of concerns around the social impact of energy cropping, there can also be significant social impacts from energy cropping in developed countries. These impacts may be positive or negative, depending on the context and the perspective from which the impacts are viewed. Energy cropping may help to maintain existing social structures and patterns, for example by helping farmers to find alternative income that keeps them on their land and creating jobs that keep people in local towns. In other cases, energy cropping may hasten demographic change through land changing hands and jobs shifting from one location to another.

Surveys and interviews with landholders and other community members can help to identify the factors that may help to make energy crops socially acceptable in different contexts. For example, in a study of landholders interested in growing mallee eucalypts near Condobolin in the central west of New South Wales, Australia, economic factors such as diversifying and improving the consistency of income were the most commonly cited potential benefits of growing mallee (Figure 6.2). Many of the factors classed as social also had a strong economic component, such as local employment and business opportunities, but the focus was more on the benefit for the local community rather than for the landholder specifically (Baumber et al., 2011).

Social and economic factors are often interlinked within the underlying value systems of landholders and other community stakeholders, and can have complex effects on attitudes towards energy cropping. In the Condobolin study, the most widely-favoured business model for growing mallee was a community-based option in which locally-owned businesses would coordinate planting, harvesting and processing (Baumber et al., 2011). The advantages cited for this model were greater control for landholders over their land and higher potential returns, as

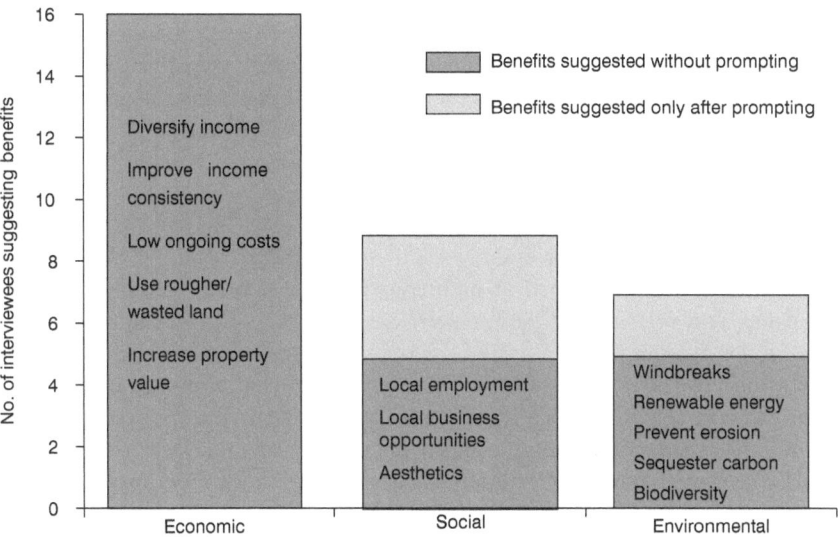

Figure 6.2 Number of interviewees identifying potential economic, social and environmental benefits of energy cropping around Condobolin. Most commonly cited benefits are listed in the each column. The light-shaded areas represent interviewees who suggested benefits only after prompting (i.e. "you've identified some economic benefits, can you see any social or environmental benefits?)

Source: Baumber et al. (2011)

well as greater flow-on benefits for the broader community. There was also some support for a different model in which an outside investor such as an energy company would contract landholders to produce mallee as a bioenergy feedstock, with the benefits of this being less risk for landholders, more stable long-term contracts, lower management requirements and access to bigger markets. Interestingly, an option which attracted no support among those interviewed, and was indeed strongly criticised by three-quarters of interviewees, was growing mallee solely for carbon sequestration under contracts that would require the mallee to be maintained unharvested for 100 years. Some of the views expressed by landholders about this option included:

- "You lose control of your land for 99 years."
- "It's inhibitive, it devalues the land."
- "If it was twice or three times [current returns] or whatever my trigger would be – people sell their souls at a price don't they? But you would take a lesser price I suppose for renewable energy – being involved in that sort of market."

These views were a direct response to the fact that some neighbouring landholders had already entered into these kinds of carbon sequestration contracts involving mallee eucalypts. Their decision to grow mallee for carbon clearly clashed with the values of some of their neighbours, who saw it as a form of land retirement which was contributing to local declines in farm output, associated economic activity, employment and population. In contrast, the idea of growing mallee as a short-rotation energy crop was consistent with the interviewees' core values that farmland should be used for productive purposes and that landholders should retain control over how the land is used. When asked if there should be limits on the amount of land used for energy cropping, a small minority of interviewees argued that some land should be retained for food production, but most felt that the level of energy cropping should be a matter for individual landholders and would be dictated by economic viability.

Social analysis undertaken at different locations may reveal differences in the values and priorities of local stakeholders. For example, a comparison case study undertaken in parallel to the Condobolin mallee study revealed a stronger level of interest in the potential environmental benefits of energy crops, such as windbreaks, habitat for biodiversity and salinity mitigation (Figure 6.3). This second case study was undertaken in a different region of NSW, the Central Tablelands, which has higher rainfall, is closer to the Sydney metropolitan area, experiences colder and windier conditions (hence the concern about shelter for stock) and has a higher incidence of salinity problems.

Economic factors may have been less influential in the Central Tablelands compared to Condobolin due to there being a higher proportion of landholders who were not reliant on their land for income, less severe impacts from recent droughts on traditional farming activities, a more diversified local economy and a more stable rural population. The level of knowledge about energy cropping was also much lower in the Central Tablelands. Radiata pine is commonly grown for timber there, but surveyed landholders had limited knowledge of growing eucalypts commercially and no experience at all with energy cropping. The eucalypt species most likely to be suited to energy cropping in the tablelands were larger trees like Tasmanian Blue Gum rather than mallee.

Many of the same issues highlighted by these Australian case studies also appear in the results of a UK study into community attitudes towards energy cropping (Dockerty et al., 2012). This study looked at both SRC tree crops and miscanthus in the East Midlands and the South West of England and had a stronger focus on the aesthetic impacts of energy crop expansion on the landscape. As with the NSW case studies, many participants had little direct experience with energy crops, so photos and computer-generated images were used to convey the landscape impacts of SRC and miscanthus crops. Generally, the results of the questionnaire, focus groups and interviews showed a high level of community acceptance of SRC and miscanthus crops, but a number of issues were raised by participants. Perceived benefits of energy crops included possible improvements in landscape aesthetics, as well as local processing opportunities and use of renewable energy within the community (e.g. school heating). Perceived risks included loss of landscape amenity, increased heavy vehicle movements and concerns about "food versus fuel".

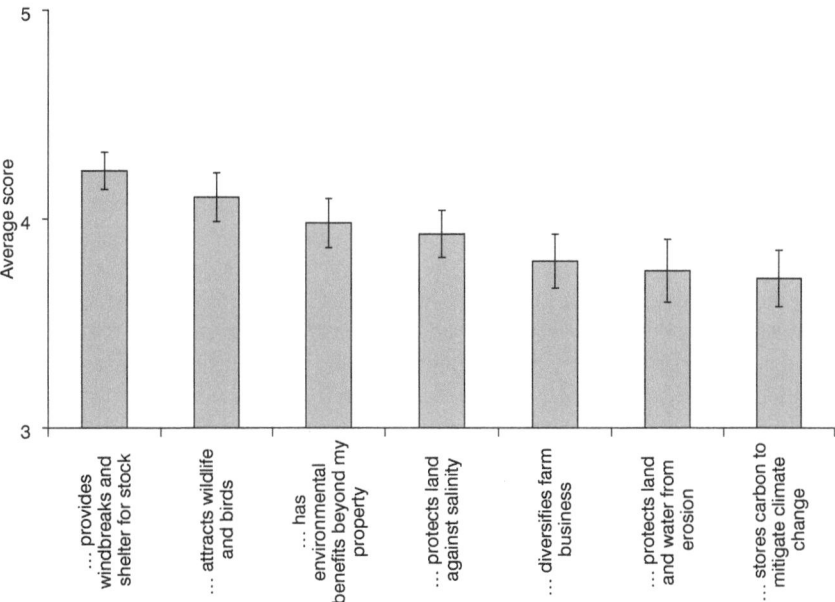

Figure 6.3 Highest ranking economic, social and environmental benefits among surveyed stakeholders in the Central Tablelands of NSW. Scores are based on the level of agreement with statements about potential benefits (Likert scale from 0 to 5)

Source: Baumber et al. (2012)

Predicting the responses of a local community to a large-scale expansion of energy crops can be difficult before such changes have actually taken place. However, it may be possible to learn some lessons from community responses to plantation expansion that has occurred for other purposes, such as timber or pulpwood production. There are a number of examples from Australia and New Zealand where rapid rates of plantation expansion have met with resistance in some rural communities due to factors such as loss of good agricultural land, competition for water, use of chemicals, visual amenity and population decline (e.g. Tonts and Schirmer, 2005; Parsons et al., 2007; Wilderness Society Tasmania, 2013). Based on survey data from Western Australia and Tasmania, Williams (2009a, 2009b) suggests that plantations are more likely to be accepted if they do not take up the highest-quality land, are in areas where competition for water is not high, are on land previously used for other plantations, occupy only part of a property, involve local processing and are owned by farmers rather than outside investors. Many of these factors are similar to those highlighted by the NSW and UK energy cropping studies, although the relative importance of different factors is likely to vary depending on the context and the nature of the energy cropping system.

Policy options

Many of the key policy options for preventing land grabs and promoting socio-economic benefits from energy cropping will be familiar from the preceding chapters on deforestation, ecological restoration and food security. The relevant measures can be grouped into four main categories: regulatory measures applied at the point of feedstock production, supply-chain interventions such as sustainability standards, market-based instruments such as subsidies or mandates, and direct support to assist with energy crop establishment (e.g. grants, low-interest loans or research and development funding).

Regulatory measures applied at the point of feedstock production represent the most direct means of protecting land rights and ensuring fair treatment of farmers and workers. However, there can be significant obstacles to the effective protection of land tenure, as highlighted in the examples from Indonesia, the Philippines and Madagascar discussed earlier in this chapter. In many cases, traditional land tenure is not fully recognised under national laws. Even where regulations have been introduced to address this, there may be issues with corruption, a lack of resources for enforcement and disputes over who has a right to represent traditional landholders in negotiations.

The Food and Agriculture Organization of the United Nations (FAO) has produced a set of Voluntary Guidelines on the Responsible Governance of Tenure (FAO, 2012). These are aimed at national governments and provide guidance on the establishment of appropriate legal structures and institutional arrangements to protect land tenure. These guidelines highlight the need for appropriate consultation to identify rights holders, equitable arrangements that protect the rights of vulnerable groups such as women and indigenous people, adequate resources to enforce legal rights and alternatives to forced eviction in cases where legal rights cannot be established.

Another UN agency, the International Fund for Agricultural Development (IFAD) has also been active on issues of land tenure. This includes providing policy advice on improving security of tenure (IFAD, 2008) and developing a set of land access indicators that rank countries from 1-6 based on the security of land access for the rural poor (IFAD, 2003). However, accurately measuring the security of land tenure at a global scale has proven challenging, with IFAD's land access indicators limited by a lack of data for many countries. Other methodologies have also been put forward by different organisations, with the International Land Indicator Initiative set up in 2012 in an attempt to harmonise these competing approaches. The upcoming release of the Sustainable Development Goals (SDGs) by the UN in September 2015 presents another opportunity to progress this goal, with the International Land Coalition (ILC) and other groups proposing a new set of indicators based on the percentage of women, men, indigenous groups and local communities who either have legal documentation of their rights or perceive their rights to be recognised (ILC, 2015).

Aside from the efforts of inter-governmental and non-government organisations to improve the legal recognition of land rights for local people, there is also a role to be played by industry participants through exerting pressure on suppliers. As highlighted earlier in the chapter, sustainability standards such as those of the SAN, FSC or RSB are one way of doing this. A key advantage of these approaches are that they do not need to wait for the producing country's laws and regulations to be re-written or for national governments to provide the resources required for effective enforcement. However, a limitation is that these standards are voluntary, and thus their effectiveness is dependent on the level of consumer concern about particular issues and the degree to which suppliers value the market segment represented by concerned consumers. As discussed earlier in the chapter, there are good reasons to doubt whether biofuels will ever attract the same level of interest from consumers interested in "fair trade" that food products do. This is due to the lack of direct, intimate contact with biofuel products, their relatively recent introduction to many consumers and the availability of alternatives such as fossil fuels.

One option for enhancing the effectiveness of sustainability standards is for more countries to follow the lead of the EU and make compliance with such standards compulsory for fuel suppliers who wish their biofuels to be counted against national renewable energy targets. However, the EU requirements currently have limited value in relation to land rights, workers' rights and other social factors, focusing instead on environmental factors such as greenhouse gas savings and deforestation. Some standards organisations seeking approval by the EU, such as the RSB, may seek to piggy-back social criteria on top of the required criteria relating to greenhouse gas savings and forest protection, but other EU-

Box 6.1 **Promoting the social benefits of energy cropping in Brazil**

Brazil's national programme for biodiesel use and production (PNNB) began in 2004 and employs market-based incentives to encourage the production of biodiesel from oilseed crops. The PNNB followed on from Brazil's more well-known ethanol programme, PROALCOOL, and employs a similar approach based on fuel tax exemptions and a fuel blending mandate for biodiesel (initially at 2% before rising progressively to 7% in 2014). Notably, the Brazilian government chose to incorporate social objectives into its biodiesel programme by certifying certain types of biodiesel as "Combustível Social", or social fuel. These social fuels, which had to be produced from feedstocks grown by family farmers under the National Programme for the Strengthening of Family Agriculture (PRONAF), were granted a 67.9 per cent reduction in fuel tributes (taxes), increasing to 100 per cent if the biodiesel feedstocks were produced in Brazil's poorest regions in the north and north-east of the country (Pousa et al., 2007).

In its early stages, the effectiveness of the PNNB programme in enhancing social outcomes for poor family farmers was questioned (e.g. Romeiro, 2006; Pousa et al., 2007). Small farmers, mostly growing castor bean crops, struggled to compete with industrial-scale soybean producers and suffered from low prices, low yields and unstable relationships with biofuel producers. However, a major turnaround occurred after reforms were made in 2008/9 and Petrobrás (Brazil's state-owned oil company) became actively involved the social fuel market. Lima (2012) highlights five key changes made by Petrobrás that helped to bring about a change in the fortunes of smallholders under the PNNB:

1 Higher-quality seeds were provided to smallholders and other crops such as sunflower were explored.
2 Petrobrás worked with smallholder cooperatives to improve their technical and organisational capabilities.
3 Petrobrás purchased feedstock at above-market prices.
4 Integrated production of food and biodiesel feedstocks was promoted over biofuel-only monocultures to reduce the reliance of landholders on a single market.
5 Petrobrás required contracts with smallholders to be co-signed by a local social movement to verify their fairness.

Following the reforms to the PNNB, the number of smallholders participating in the programme quadrupled between 2008 and 2010 and there was a measurable rise in the reported satisfaction levels of participating smallholders. While Lima (2012) argues that some problems with the PNNB remain, such as its inability to change underlying patterns of inequality in income and land ownership, the reforms made to the programme highlight how market-based incentives that are combined with more specific measures tailored to landholder needs can create opportunities for socio-economic advancement.

approved standards have much lower coverage of social issues (German and Schoneveld, 2011). The EU and other importers of biofuel feedstocks face a major barrier to increasing their focus on social issues under sustainability criteria due to the rules of the World Trade Organization, which prevent discrimination between products based on the treatment of workers and local people (Charnovitz et al., 2008).

Regulatory measures and sustainability standards represent interventions at different points in the biofuel supply chain, but they are both aimed primarily at preventing negative outcomes, such as land-grabs or worker exploitation. In order to actively promote energy crops that offer social benefits, other measures are likely to be required. One option is to offer support directly to interested landholders through financial aid (e.g. start-up grants, tax offsets, low-interest loans) or through the extension of research findings and technical assistance. Another option is to employ market-based instruments that incentivise energy cropping, such as tax breaks, biofuel mandates, renewable energy targets or feed-in tariffs, and to tailor these programmes to preference forms of energy cropping that offer social benefits.

Box 6.1 presents an example where biofuel incentives have been tailored to promote socio-economic outcomes for smallholders under Brazil's national programme for biodiesel use and production (PNNB). This example highlights both the potential to engage smallholder farmers in energy cropping and the challenges in doing so. Smallholders in many countries have experience growing crops that can be used for bioenergy, including sugarcane, oil palm, soybean and jatropha (Hilhorst, 2014). However, the viability of these smallholder options and their socio-economic impacts are often dependent on relationships with crop buyers and processing facilities and how the costs, benefits and risks of production are divided under supply contracts.

The example of the PNNB in Brazil highlights the need to complement market-based incentives such as biofuel mandates with more targeted measures such as technical assistance and legal advice to ensure that the design of the programme aligns with landholder goals and values. This need for integrated policy measures is also supported by the results of the policy analysis undertaken for the study of mallee cropping in the Australian state of NSW discussed earlier in the chapter. In that study, landholders were asked to rank four potential support mechanisms that could assist with the development of a mallee cropping industry:

- establishment support (including both financial and non-financial support);
- market development (e.g. research and development to identify new products from mallee);
- price support (e.g. through renewable energy mandates or carbon pricing); and
- payments for ecosystem services (e.g. payments based on increases in sequestered carbon or habitat for biodiversity).

Of these options, the most favoured by far was establishment support (see Figure 6.4).

Follow-up questions to the interviewees at Condobolin revealed that the need for establishment support was largely financial, such as low interest loans (which landholders were familiar with due to their use in other agricultural industries). This result was mirrored in the sister study undertaken further east in the NSW Central Tablelands, where financial support with establishment costs was ranked first, followed by knowledge support, industry development, payments for ecosystem services and price support for the harvested products (Baumber et al., 2012).

An interesting contrast was observed in the Condobolin case study between the strong landholder preference for establishment support and the focus of certain government and industry interviewees on price support mechanisms, such as mandates for renewable energy and carbon pricing. While these stakeholders saw these kinds of market-based incentives as essential to the establishment of a viable mallee cropping industry, many landholders were concerned about the risks

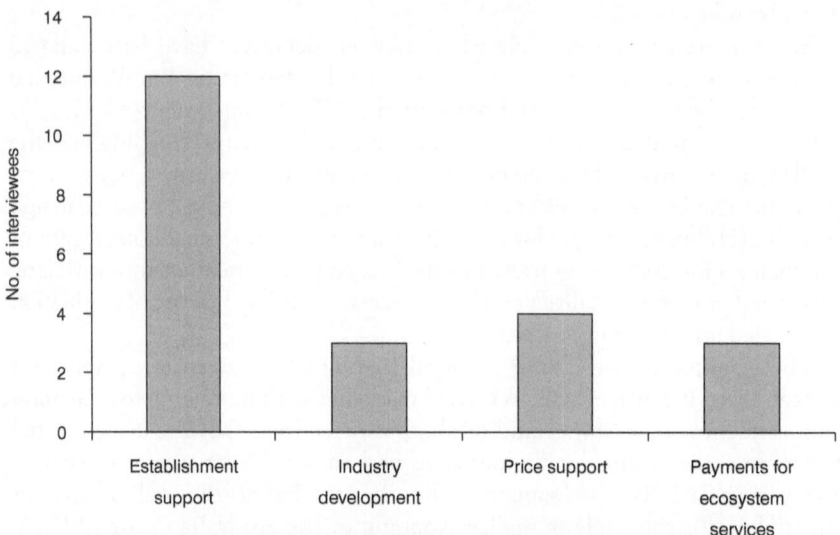

Figure 6.4 Support mechanisms ranked first or equal first by interviewees in the Condobolin mallee cropping study

Source: Baumber et al. (2011)

involved in having to take on the upfront establishment costs themselves while there was so much political uncertainty surrounding the future of carbon pricing and the Renewable Energy Target in Australia. Landholders were also sceptical about price support measures due to witnessing to the progressive removal of price support for Australian agriculture over several decades (e.g. the floor-price for wool, the fixed exchange rate and the single export desk for wheat).

Despite the very different social contexts in Australia and Brazil, the NSW eucalypt cropping and Brazilian biodiesel examples show some distinct similarities. In the case of Brazil's PNNB, the government's initial approach was to stimulate demand for biodiesel feedstocks from poor farmers via a broad-scale market intervention, while the farmers themselves were in need of much more direct support relating to seeds, technical skills and the negotiation of contracts. In Australia, there has also been a focus on national-scale market-based policy measures such as fuel tax rebates, carbon pricing and the Renewable Energy Target to promote energy cropping as a form of renewable energy production. However, as the results of the NSW social analysis show, landholders are likely to view reliance on these market-based drivers as risky, especially for untested crops with significant establishment costs. As in Brazil, support with upfront costs and technical assistance are likely to be important complements to market-based measures.

Conclusion

As with the issues of food security and deforestation discussed in preceding chapters, it is clear that energy cropping is only one element in the overall matrix of land uses that impact on the social wellbeing of local communities across the globe. However, energy cropping does have the potential to exert a notable and specific influence on issues of land rights and socio-economic development in certain contexts. In some locations, the demand for biofuel feedstocks may be exerting a disproportionately high pressure on local communities in the form of large-scale land acquisition, particularly as energy cropping represents a relatively new land use option experiencing rapid growth due to strong support from governments in developed countries. In such cases, it is prudent to look at whether changes to the way that biofuels are promoted could help to overcome these problems, such as the addition of social sustainability criteria to government support schemes, or increased engagement with inter-governmental processes aimed at enhancing the protection of land rights for local and indigenous people.

While energy cropping may be capable of contributing to global problems around land-grabs and worker exploitation, it is important to remember that it also has the capacity to contribute to socio-economic opportunities, such as increases in rural incomes, job creation and enhancement of landholder goals relating to landscape management and aesthetics. Furthermore, because of the rapid innovation that has occurred in bioenergy policy around the world, energy cropping could also play a leading role in identifying ways to promote positive social impacts while guarding against risks. Two key ways of doing this that have been highlighted in this chapter are the careful integration of sustainability standards into biofuel support schemes and the tailoring of market-based incentives to direct support to those most in need of socio-economic opportunities.

The economic opportunities that energy cropping can create are further explored in the following chapter on the economics of energy cropping, before the various policy ideas from each chapter are brought together in the search for a sustainable way forward in Chapters 8 and 9.

References

Anseeuw, W., Wily, L. A., Cotula, L. and Taylor, M. (2012) *Land Rights and the Rush for Land: Findings of the Global Commercial Pressures on Land Research Project*, International Land Coalition, Rome.

Baumber, A. P., Merson, J., Ampt, P. and Diesendorf, M. (2011) "The adoption of short-rotation energy cropping as a new land use option in the New South Wales Central West", *Rural Society*, 20: 266–79.

Baumber, A., Rammelt, C., Ampt, P. and Merson, J. (2012) *Bioenergy from Native Agroforestry: Planning for a Regional Industry in the NSW Central Tablelands*, Rural Industries Research and Development Corporation, Canberra.

Carbon Market Watch (2010) "Plantar – pig iron project, Brazil", http://carbonmarketwatch. org/campaigns-issues/plantar-pig-iron-project-brazil (accessed 10 July 2015).

Charnovitz, S., Earley, J. and Howse, R. (2008) *An Examination of Social Standards in Biofuels Sustainability Criteria*, International Food and Agricultural Trade Policy Council, Washington, DC.

Colchester, M. (2011) *Palm Oil and Indigenous Peoples in South East Asia*, International Land Coalition, Rome. 6p.

Dockerty, T., Appleton, K. and Lovett, A. (2012) "Public opinion on energy crops in the landscape: considerations for the expansion of renewable energy from biomass", *Journal of Environmental Planning and Management*, 55: 1134–58.

Eide, A. (2008) *The Right to Food and the Impact of Biofuels (Agrofuels): Advanced Copy*, Food and Agriculture Organization of the United Nations, Rome.

Elbehri, A., Segerstedt, A. and Liu, P. (2013) *Biofuels and the Sustainability Challenge: A Global Assessment of Sustainability Issues, Trends and Policies for Biofuels and Related Feedstocks*, Food and Agriculture Organization of the United Nations, Rome.

European Commission (2012) *Proposal for a Directive of the European Parliament and of the Council Amending Directive 98/70/EC Relating to the Quality of Petrol and Diesel Fuels and Amending Directive 2009/28/EC on the Promotion of the Use of Energy from Renewable Sources*, European Commission, Brussels.

Fairtrade Foundation (2015) "What product categories does fairtrade certify?", www. fairtrade.org.uk/en/what-is-fairtrade/faqs (accessed 15 May 2015).

FAO (2012) *Voluntary Guidelines on the Responsible Governance of Tenure of Land, Fisheries and Forests in the Context of National Food Security*, Food and Agriculture Organization of the United Nations, Rome.

Fortin, E. and Richardson, B. (2013) *Certification Schemes and the Governance of Land: Enforcing Standards or Enabling Scrutiny?* Routledge, Abingdon.

Friends of the Earth (2010) *Jatropha: Money Doesn't Grow on Trees*, Friends of the Earth, London.

German, L. and Schoneveld, G. (2011) *Social Sustainability of EU-Approved Voluntary Schemes for Biofuels: Implications for Rural Livelihoods*, Center for International Forestry Research (CIFOR), Bogor.

Hilhorst, T. (2014) "How land rights influence the social impacts of biomass production", in Langeveld, J. W. A., Dixon, J. and Keulen, H. v. (eds), *Biofuel Cropping Systems: Carbon, land and food*, Routledge, Abingdon, 51–62.

IFAD (2003) *The Structure and Operation of a Performance-Based Allocation System for IFAD*, International Fund for Agricultural Development, Rome.

IFAD (2008) *Improving Access to Land and Tenure Security*, International Fund for Agricultural Development, Rome.

ILC (2015) *Land Rights Indicators in the Post-2015 SDGs: Recommendations for Inter-Agency Expert Group and Other Policymakers*, International Land Coalition, Rome.

Land Matrix (2015) *Land Matrix: Whole Database*, Land Matrix, http://landmatrix.org/en/get-the-detail/ (accessed 11 March 2015).

Lima, M. B. (2012) *An Institutional Analysis of Biofuel Policies and their Social Implications Lessons from Brazil, India and Indonesia*, United Nations Research Institute for Social Development, Geneva.

McMichael, P. (2010) "Agrofuels in the food regime", *Journal of Peasant Studies*, 37: 609–29.

Parsons, M., Frakes, I. and Gerrand, A. (2007) *Plantations and Water Use*, Bureau of Rural Sciences, Canberra.

Pousa, G. P. A. G., Santos, A. L. F. and Suarez, P. A. Z. (2007) "History and policy of biodiesel in Brazil", *Energy Policy*, 35: 5393–8.

Ratsialonana, R. A., Ramarojohn, L., Burnod, P. and Teyssier, A. (2011) *After Daewoo? Current Status and Perspectives of Large-Scale Land Acquisitions in Madagascar*, International Land Coalition, Rome.

Ravanera, R. and Gorra, V. (2011) *Commercial Pressures on Land in Asia: An Overview*, International Land Coalition, Rome.

Romeiro, A. R. (2006) *Biofuels in Brazil: A Prospective Option Against Deforestation, Income Concentration and Regional Disparities*, Netherlands Environmental Assessment Agency, Utrecht.

Sheil, D., Casson, A., Meijaard, E., Noordwijk, M. v., Gaskell, J., Sunderland-Groves, J., Wertz, K. and Kanninen, M. (2009) *The Impacts and Opportunities of Oil Palm in Southeast Asia: What Do We Know and What Do We Need to Know?*, Center for International Forestry Research, Bogor.

Tonts, M. and Schirmer, J. (2005) "Managing Social Conflict in the Tree Plantation Industry: Growing Consensus or Deepening Divisions?", in Cryle, D. and Hillier, J. (eds), *Consent and Consensus: Politics, Media and Governance in Twentieth Century Australia*, API Network, Perth, 275–96.

Wilderness Society Tasmania (2013) "Statement of principles a great start – but there's more to do", www.wilderness.org.au/regions/tasmania/statement-of-principle-a-great-start-but-theres-more-to-do/?searchterm=%20plantations%20tasmania (accessed 26 February 2013).

Williams, K. (2009a) *Community Attitudes to Plantations: Survey of the Views of Residents of South-west Western Australia 2008*, CRC Forestry, Hobart.

Williams, K. (2009b) *Community Attitudes to Plantations: Survey of the Views of Residents of Tasmania 2008*, CRC Forestry, Hobart.

Zoomers, A. (2010) "Globalisation and the foreignisation of space: seven processes driving the current global land grab", *Journal of Peasant Studies*, 37: 429–447.

Chapter 7

The economics of energy cropping

According to the "three-pillar" model of sustainability discussed in Chapter 1, economics constitutes one of the three pillars, alongside social and environmental sustainability. However, when sustainable bioenergy or sustainable biofuels are discussed in the popular media or in campaigns by NGOs, economics can sometimes seem like the forgotten pillar. To those outside the bioenergy sector, discourses around "making biofuels sustainable" or "preventing unsustainable impacts from energy cropping" tend to revolve around environmental and social concerns such as deforestation, greenhouse gas emissions, food security and land rights. In this context, energy cropping is often viewed as a threatening process that needs to be controlled, with the question of whether or not it is economically viable being a secondary consideration.

While many outside the bioenergy industry may not give economic viability a great deal of thought, for those working directly with energy crops it is critically important. Sustainability standards developed with strong industry input, such as those of the Roundtable on Sustainable Biomaterials (RSB) and the Forest Stewardship Council (FSC), tend to require producers to have a plan that explicitly considers economic viability. In contrast, economic viability is much less prominent in the standards of the Sustainable Agriculture Network (part of the Rainforest Alliance), which require a social and environmental management system but not an economic one. However, even for stakeholders whose primary concerns are social or environmental, there are good reasons to take an interest in the economics of energy cropping. As highlighted in the preceding chapters, well-designed energy cropping systems can be a force for positive change through climate change mitigation, ecological restoration and social development – but only if they are economically viable.

Chapter 2 highlighted the potential for energy crops to mitigate global warming by replacing fossil fuels in the transportation and electricity sectors. Governments across the world have introduced policy incentives such as mandates, subsidies and feed-in tariffs to further this aim. Each of these policies requires a detailed and up-to-date understanding of the economics of bioenergy production within the jurisdiction they cover, as well as the supply chains that feed that production. Furthermore, if new energy crops are to become widespread, such as switchgrass, eucalypts and other cellulosic crops, economic analysis is required to determine the differences in production costs between these crops and the first-generation feedstocks that currently dominate the biofuel market.

Chapter 4 introduced the notion of conservation through sustainable use (CSU), which suggests that commercial use may actually be preferable to no use at all in some cases, especially when the use activity creates an incentive for local people to value and protect the ecosystems that support it. The Mediterranean cork forests and the damar agroforests of Indonesia were cited as two such ecosystems where commercial use provides an incentive for conservation – and it is conceivable that energy cropping systems involving willow in Europe or mallee in Western Australia could come to be seen in a similar light in years to come. When viewed through the lens of CSU, economic sustainability is not simply about generating profits for particular stakeholders, but is also about the creation and maintenance of incentives for sustainable land use.

For those with an interest in promoting energy crops for their environmental and social benefits, the development of economically viable production models can be a daunting task, especially where trade-offs are required between production efficiency and benefits for ecosystem health. Jatropha, as discussed in Chapter 4, provides an example of an energy crop that has failed to live up to its early hype and has generated a backlash due to a lack of economic viability. The credibility of jatropha today may have been stronger had there been less hype about the idea of an inedible plant that would grow well in poor conditions, combat desertification and enhance food security. Instead, a greater focus on first developing economically viable business models which could then be adapted to areas with environmental challenges may have created a different reputation for jatropha today. There are important lessons here for the development of energy crops that may be able to assist in the restoration and protection of degraded or vulnerable landscapes, such as mallee eucalypts, willow and switchgrass.

This chapter discusses the economic factors affecting the production of both well-established and emerging bioenergy cropping options, with a focus on two main categories of energy crops – established biofuel crops (i.e. for first-generation biofuels) and cellulosic (i.e. tree or grass) crops used for electricity, heat and increasingly for advanced biofuels. Biofuels from algae or waste feedstocks are not considered due to their lesser capacity to impact on land use patterns, either positively or negatively.

The statistics cited are drawn from recent studies into the economics of bioenergy production, but they are inevitably subject to change and uncertainty, especially for emerging crops and advanced conversion processes. Apart from advances in bioenergy technologies and cropping systems, the viability of established and emerging energy crops is also affected by volatility around key economic parameters such as global oil prices, exchange rates, agricultural commodity prices and policy support programmes.

Established biofuel crops

The most important first-generation biofuel feedstocks are corn and sugarcane, used to produce ethanol, and a variety of oil-bearing crops used to produce biodiesel. A range of countries have established viable production systems for these crops, most notably the USA for corn ethanol, Brazil for sugarcane ethanol and various EU countries for biodiesel. The viability of each cropping system depends on the balance between cost of production (e.g. feedstock production, transport, processing and conversion) and the income received from a combination of biofuel sales and government support.

Market prices for biofuels are heavily influenced by global oil prices, but government policy also plays an important role. Support for biofuels may be aimed at influencing the market price for biofuels, such as through mandates that boost demand by requiring fuel suppliers to blend biofuels with gasoline or diesel. Alternatively, support may be provided in a manner that is additional to the market prices that biofuel suppliers receive, such as when fuel taxes collected at the point of sale are rebated to biofuel suppliers (as has been practiced in Australia for example). Other policy interventions may preference certain biofuels over others, such as tariffs that provide domestic biofuels with an advantage over their imported counterparts.

Corn grown for ethanol in the United States is one of the most successful energy crops established to date and the US Federal Government has employed all three of the policy approaches described above to enable the rapid growth of the industry. Until the end of 2011, a tax credit of 45 cents per gallon was paid to US-based ethanol fuel blenders and a tariff of 54 cents per gallon was applied to imported ethanol to reduce competition, particularly from the more-established Brazilian industry. Blending mandates were introduced in 2006 under the Renewable Fuel Standard (RFS), with progressively rising requirements for overall supply of renewable fuel. However, the focus of the RFS has been shifting away from corn ethanol towards advanced biofuels under the 2010 changes discussed in Chapter 2 (Environmental Protection Agency, 2010).

The combined impact of tax credits, tariffs, blending mandates and research and development funding has been to enable the US ethanol industry to undergo a major expansion of production, from less than 100 million gallons in 1981 to more than 14 billion gallons (53 billion litres) in 2014 (Energy Information Administration, 2015). The key impact of these measures was not so much the direct subsidies provided to ethanol producers, but rather the role they played in enabling the industry to obtain the critical mass needed to pass through a learning curve and bring down production costs. Hettinga et al. (2009) estimate that processing costs for corn ethanol in the US declined by 45 per cent between 1983 and 2005 to around US$0.13 per litre (US$130 per m³), with cost reductions achieved in the use of enzymes, labour and energy (Figure 7.1). This was complemented by a 62 per cent decline in corn production costs between 1975 and 2005, mostly due to increasing corn yields and farm sizes. Analysis by

Chen and Khanna (2012) found that the reduction in processing costs in the US corn ethanol sector was mostly due to "learning by doing" within the industry rather than external factors such as technological breakthroughs made elsewhere or changes to the cost of energy or labour.

Processing costs for US corn ethanol have not declined substantially since 2005 and are highly dependent on the fluctuating prices of corn used as feedstock and natural gas used for processing. The US Energy Information Administration estimated that fixed processing costs for 2012 were US$0.35 per gallon (US$0.09 per litre), not including the cost of natural gas or corn (Energy Information Administration, 2012). The processing costs specifically attributable to ethanol can be lowered to US$0.07 per litre by accounting for the fact that ethanol only makes up around 75 per cent of the output from an ethanol plant (the other 25% being dried distillers grain used as animal feed). This 2012 figure of US$0.07 per litre is similar to the US$0.06 per litre (i.e. US$60 per m³) for the non-energy processing costs shown in Figure 7.1 for 2005. Factoring in natural gas costs, corn costs and sales of dried distillers grain, the overall production cost for US corn ethanol was around US$2 per gallon (US$0.53 per litre) in early 2012 (Energy Information Administration, 2012), with the fluctuating cost of corn feedstock accounting for around two-thirds of the overall production costs.

Notwithstanding the levelling off of production costs in recent years, the success in reducing costs for the US corn ethanol industry from the 1980s to 2000s helped make it politically viable to introduce a blending mandate in 2006,

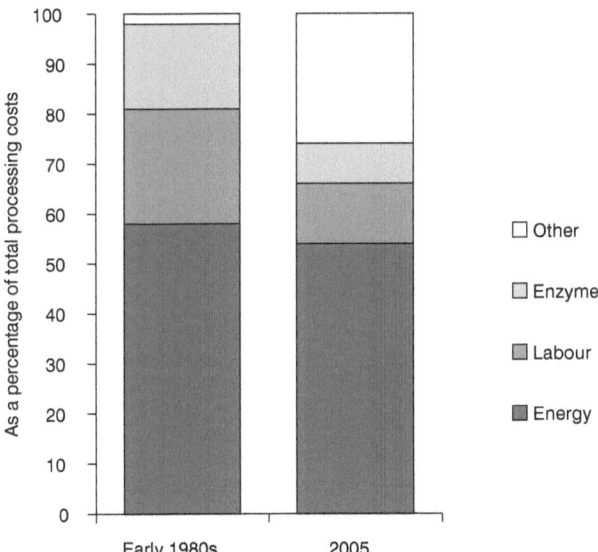

Figure 7.1 Processing costs for US corn ethanol for early 1980s and 2005
Source: Hettinga et al. (2009)
Note: Processing costs are measured in 2005 US dollars. 1 m³ = 1000 litres.

which in turn gave the industry the certainty of demand that it needed to invest in new production facilities. By 2011, a number of factors were creating resistance to further assistance for the corn ethanol industry, including political pressure to reign in US Government spending and the negative publicity generated by debates around "food versus fuel" and indirect land use change. However, the strongest argument for removing the tax credit and tariff arrangements was that US ethanol producers had become globally competitive without this support, as demonstrated by the country's transition in 2010 from a net ethanol importer to a net exporter.

External factors such as high global oil prices from 2010 to 2014 helped to make US ethanol attractive and strong demand for dried distillers grain boosted the viability of ethanol production. However, the efficiencies that had developed over many years while the industry was supported clearly helped with keeping costs down. The removal of the tax credit and tariff did see a small temporary increase in imports from Brazil in early 2012, but this soon subsided and the US continued to be a major ethanol exporter with producer margins remaining positive (Energy Information Administration, 2012). Lower oil prices in 2014–15 have had a dampening effect on ethanol prices, but this has also been accompanied by a fall in corn prices, which has helped to maintain margins for ethanol producers.

The history of ethanol economics in Brazil, the world's second largest producer, shares some similarities with the US story but has followed a different trajectory over time. Ethanol from sugarcane began being heavily promoted in Brazil in the 1970s under the PROALCOOL programme, which has helped to stimulate production and drive down costs over the past four decades and has become a model for biofuel support programmes in many other countries. This programme commenced in 1974 following the global oil crisis that impacted on the availability and price of imported petroleum products. At its commencement, ethanol was promoted by requiring purchases by the state-owned oil company (Petrobrás), setting fixed prices for ethanol and gasoline, and providing low-interest loans to ethanol producers. Over time, mandatory blending became a key component, first set at 4.5 per cent in 1977 before rising to 22 per cent by 1993 and settling on a range of 18–25 per cent in 2007 that takes account of supply shortages and high prices that occur during the sugarcane harvest season (Langeveld and Quist-Wessel, 2014). Ethanol is also given an advantage over oil-based products through fuel tax exemptions and lower vehicle taxes applied to vehicles that are designed to use ethanol only or any blend of ethanol and gasoline (flex-fuel vehicles).

From the 1970s to the mid-2000s, Brazil was the world's largest and lowest-cost ethanol producer. However, it was overtaken by the US in terms of production in 2006 and its status as the lowest-cost producer is also under threat. Méjean and Hope (2010) found that production costs for Brazilian sugarcane ethanol were around 24 per cent lower than US corn ethanol in 2006–8, with Langeveld and Quist-Wessel (2014) finding that this cost advantage persisted in 2010. However, after factoring in the value of ethanol co-products (dried distillers grain) and transport costs to bring ethanol to the US market, Brazilian ethanol becomes higher in cost than US corn ethanol (Méjean and Hope, 2010), a key factor in the US removing its tariff on imported ethanol at the end of 2011. The difference between Brazilian and US ethanol costs at any point in time is variable, being dependent on the fluctuating prices for the key inputs and co-products, as well as the exchange rate between the US Dollar and the Brazilian Real.

In Europe, the dominant biofuel is biodiesel, which has risen dramatically in production from 800 million litres in 2000 to over 10 billion litres in 2010 (Langeveld et al., 2014). Both biodiesel and ethanol are supported by the EU's Renewable Energy Directive (RED), which sets a target of 10 per cent of transport fuels from renewable source by 2020. To meet this EU-wide target, individual member states may choose to implement various support policies, such as the Renewable Transport Fuel Obligation (RTFO) in the UK, which mandates the use of renewable fuels by fuel suppliers. In addition, biofuels are eligible for reductions or exemptions on fuel excise and receive government funding for research and development. It is also important to note that the EU targets apply to biofuel consumption rather than production, which has helped to make the EU a major target market for overseas producers of biodiesel and feedstocks (although domestic production is favoured through the application of import taxes and duties on imported biofuels).

The Global Subsidies Initiative of the International Institute for Sustainable Development estimated the overall level of support for biofuels in the EU in 2011 at €0.32–0.39 (US$0.35–0.43) per litre of biodiesel consumed and €0.15–0.21 (US$0.17–0.23) per litre of ethanol consumed (GSI-IISD, 2013). The main reason for the higher figure for biodiesel is the higher level of market price support estimated by GSI. Mandates such as the UK RTFO support the price of biofuels by creating a sub-market for biofuels within the broader transport fuel market, in which there is no direct competition with fossil fuels. In addition, taxes and duties on imported biofuels elevate biofuel prices within the EU relative to global prices.

Estimating the price gap between EU biofuels and global averages is an inexact science and is subject to a range of assumptions. The GSI analysis estimated the

Table 7.1 Price gaps between EU biofuels and global averages for 2011

Factor	Ethanol	Biodiesel
EU production (million litres)	4392	9743
EU imports (million litres)	1822	3562
EU wholesale price (€/litre)	0.58–0.63	0.83–0.90
World average price (€/litre)	0.47	0.62
Transport and distribution costs (€/litre)	0.04	0
Price gap (€/litre)	0.05–0.12	0.22–0.28

Source: GSI IISD (2013)

Note: Transport costs for ethanol assume import from Brazil, while transport costs for biodiesel are zero due to the EU being the dominant global producer. €1=$US1.11 as of mid-2015.

price gap for ethanol at €0.05–0.12 (US$0.06–0.13) per litre and for biodiesel at €0.22–0.28 (US$0.24–0.31) per litre (Table 7.1).

Despite the EU being the dominant producer of biodiesel, there are a range of factors that can make production costs there higher than in other parts of the world, including higher feedstock prices, energy and labour costs and the costs of complying with health and safety and sustainability requirements. Table 7.2 shows a range of estimates from previous studies compiled by Ong et al. (2012). Palm oil represents one of the cheapest biodiesel options in Table 7.2, with the production cost of US$0.63 for Malaysian palm oil biodiesel estimated by Ong et al. (2012) being very similar to the figure of US$0.54 cited by Langeveld et al. (2014). These production costs are well below the prevailing EU wholesale prices for biodiesel shown in Table 7.1, which highlights why there has been a major shift in the EU away from the traditionally dominant rapeseed oil towards imported palm oil as a biodiesel feedstock (Gerasimchuk and Koh, 2013).

Jatropha is another potential feedstock for biodiesel. While it has struggled to become established as a commercially viable crop in many parts of the globe, it is still considered "first-generation" in the sense that it involves a well-established conversion pathway, with oils first being extracted from jatropha seeds, followed by trans-esterification to create biodiesel. Due to the failure of many jatropha plantations established in the past decade, and the lack of clearly viable production models, it is difficult to make obtain meaningful figures about the economics of jatropha production. Soto et al. (2013) cite evidence from Ethiopia that jatropha seeds were attracting a price of US$136–161 per tonne in 2011, well short of the US$210 per tonne estimated to be the breakeven price for large-scale cultivation. Biodiesel production from jatropha was found to be more competitive with fossil diesel when grown as a fence around fields, as it avoided the costs associated with converting natural vegetation to large-scale plantations.

Table 7.2 Biodiesel production costs from selected studies

Feedstock	Plant capacity (t/year)	Location	Feedstock cost (US$/t biodiesel)	Glycerol credit (US$/t biodiesel)	Biodiesel cost (US$/ litre)	Year study published
Waste cooking oil	7260	Japan	248	0	0.58	2009
Soybean oil	8000	USA	779	380	0.78	2008
Rapeseed oil	50,000	Greece	1158		1.15	2009
Rapeseed oil	8000	Denmark	3042	2215	2.04	2010
Castor oil	8650	Brazil	1156	44.1	1.56	2010
Palm oil	36,000	Mexico	358	33.5	0.37	2010
Palm oil	1000	India	588	200	2.30	2011
Palm oil	50,000	Malaysia	568	13	0.63	2012

Source: Ong et al. (2012)

Similarly, van Eijck et al. (2012) found that biodiesel from jatropha could be competitive with fossil diesel in an east African smallholder setting, but only if family labour was used to reduce costs.

Cellulosic energy crops

As highlighted in the preceding chapters, tree or grass crops grown for their lignocellulose content (i.e. woody or fibrous component) have the potential to produce higher greenhouse gas savings, reduce competition with food production and contribute to the restoration of degraded land. Some of these crops have already become established as energy crops, particularly willow, poplar and miscanthus grown for electricity and heat in Europe. Others, such as switchgrass in the US and eucalyptus trees in Australia, are yet to be fully commercialised as energy crops, but have attracted considerable attention in terms of research and development. Due to the emerging nature of these crops, there is much uncertainty surrounding the economics of both the cropping techniques involved and the technologies for converting cellulosic biomass into liquid fuels.

The most established cellulosic energy crops at present are willow, poplar and miscanthus. These crops have proven to be economically viable for electricity generation and industrial heat, especially in Europe. El Kasmioui and Ceulemans (2012) reviewed 23 studies looking at the economics of short-rotation cropping using willow or poplar, mostly in Europe but also including the USA, Canada and Chile. They found that production costs in € per GJ (i.e. gigajoule of energy contained in the biomass) varied between 0.8 and 5 (Figure 7.2).

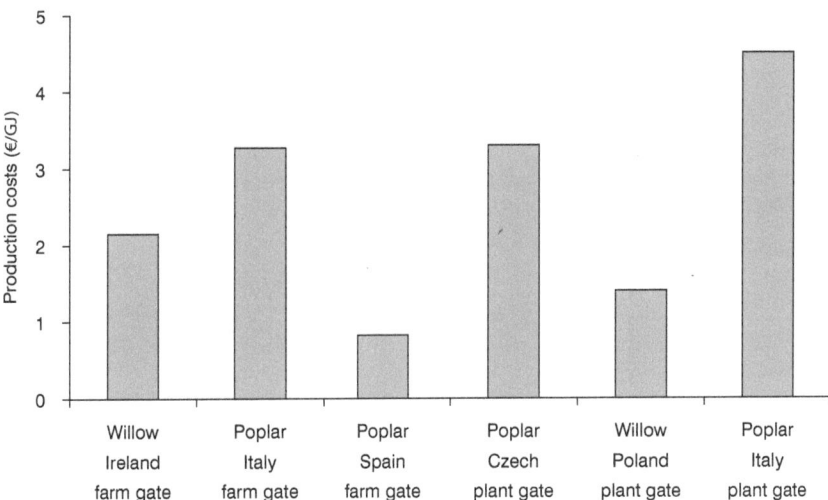

Figure 7.2 Estimated production costs for energy crops from selected studies
Source: El Kasmioui and Ceulemans (2012)
Note: A mid-range value has been used for studies that published results as a range.

While Figure 7.2 presents the results of different studies side-by-side, care should be taken when making direct comparisons, due to the differing methodologies and assumptions used. Some of the studies used a life-cycle boundary up to the farm gate only, while others included the costs of transport to the gate of the biomass plant. Any costs related to the conversion of the biomass to electricity or heat (as well as any pre-processing required) are in addition to the production costs shown in Figure 7.2.

The lower end estimates in Figure 7.2 compare favourably with coal prices in northwestern Europe of around €2.50 per GJ in 2012 (Ecofys, 2012). Energy crops can be co-fired with coal to reduce the need for new generation facilities to be built. Alternatively, standalone biomass facilities may be built with a focus on combined heat and power (CHP) rather than electricity-only. This can greatly increase the efficiency of overall energy use, as electricity generation plants often lose up to two-thirds of the fuel's energy content as waste heat. CHP is most viable where there is ready market for the heat produced, such as in a processing facility for wood products or sugar, or a nearby urban area with a centralised heating system.

The economic viability of short-rotation energy crops in Europe is highly dependent on policy measures such as farm subsidies, assistance with establishment costs, carbon pricing and renewable energy mandates. El Kasmioui and Ceulemans (2012) found that Spain and Poland were the only countries they reviewed where subsidies were of minor importance to the economic viability of energy cropping. At the EU level, energy crops are eligible for subsidies on a per hectare basis under the Common Agricultural Policy (CAP) as a form of "carbon payment", while national level support programmes vary from country to country with regard to their assistance with establishment and ongoing costs.

In addition to the support provided at the point of cultivation, the RED and EU Emissions Trading Scheme (ETS) help to make biomass fuels competitive with fossil fuels at the point of combustion. The RED targets are supported by mandates such as the UK's Renewables Obligation, which requires electricity providers to use a mandated amount of renewable generation. Obligations are met by surrendering certificates to the Government, with certain types of energy crops being eligible for more certificates per megawatt-hour (MWh) than established bioenergy technologies, like landfill gas (Table 7.3). The EU ETS also helps to make biomass more competitive by pricing the greenhouse gas emissions from coal (around €7 per tonne of carbon dioxide equivalent in early 2015).

The potential to convert woody energy crops to liquid biofuels rather than electricity opens up a range of new possibilities. Much of the work to date on cellulosic biofuels has focused on the use of waste biomass from agricultural or industrial processes, as these sources of biomass are often cheaper and more readily available than biomass from energy crops. However, if the economics of cellulosic biofuel production continues to improve and reach parity with the costs of producing fossil fuels, the supply of existing wastes may prove insufficient and energy crops may prove cheaper than many wastes that are difficult to collect efficiently (e.g. in-field agricultural residues).

Techniques to convert cellulosic biomass into ethanol generally focus on the use of enzymes to break down the lignocellulose and convert it into fermentable sugars. Thermochemical processes such as gasification, pyrolysis and Fischer–Tropsch conversion can also be applied to cellulosic biomass to create substitutes for fossil diesel. Such fuel is often termed *renewable diesel* to differentiate it from biodiesel produced through the transesterification of plant-based oils.

Stephen et al. (2013) employed recent estimates of cellulosic ethanol production costs to predict the minimum ethanol selling price (MESP) that would be required by plants situated on the east and west coasts of North America. The MESP was estimated to fall in the range US$0.75–1.00 per litre, which is substantially higher than the global average ethanol price of around US$0.50 shown previously in Table 7.1. In terms of thermochemical conversion to produce renewable diesel, recent techno-economic analyses indicate production costs of around US$1.15 per litre using gasification and US$1.57 per litre using Fischer–Tropsch conversion (Brown, 2015). As with the cellulosic ethanol analysis of Stephen et al. (2013), these renewable diesel costs are substantially higher than prevailing biodiesel prices shown in Table 7.1 (around US$0.69 per litre). This highlights why the US EPA has been forced to repeatedly revise its cellulosic biofuel requirements downwards, with the 2014 target scaled back to only 1 per cent of the target originally set in 2010 (Environmental Protection Agency, 2013).

While the costs of cellulosic biofuels are high at present and there is a lack of large-scale facilities, targeted research and learning-by-doing within the industry is likely to reduce costs over time. Cost forecasts for cellulosic fuels by the International Energy Agency suggest that production costs will be comparable

Table 7.3 "Banding" arrangements for bioenergy under the UK Renewables Obligation

Generation type	Credits per MWh
Landfill gas	0.25
Sewage gas, co-firing on non-energy crop (regular) biomass	0.5
Co-firing of energy crops Co-firing on non-energy crop (regular) biomass with CHP The use of fuels made using standard gasification or pyrolysis	1.0
Co-firing of energy crops with CHP Dedicated regular biomass	1.5
Fuels made using anaerobic digestion Advanced gasification or pyrolysis Dedicated biomass burning energy crops (with or without CHP) Dedicated regular biomass with CHP	2.0

Source: Department of Energy and Climate Change (2012)

Note: CHP = combined heat and power.

with current prices by 2030 (Ecofys, 2013). If such forecasts are achieved, this could see a number of changes, including a decline in the use of first-generation feedstocks (as cellulosic biomass would offer a cheaper alternative), a switch in focus from electricity to liquid fuels for established crops such as willow and miscanthus, and the emergence of new cropping systems that are not viable at present. New multi-purpose cropping systems may also emerge, with some components being used for food or sent to bio-refineries for the extraction of high-value chemicals, while other components are used for energy.

The aforementioned study of cellulosic ethanol production in North America (Stephen et al., 2013) highlights the potential for new cropping systems to emerge in response to demand from cellulosic fuel producers. While that study showed that none of the options were competitive with prevailing ethanol prices, it also showed that it would be cheaper for North American producers to import short-rotation eucalypts from Brazil than to use locally grown willow or poplar. The costs of shipping biomass across the ocean were more than offset by the higher yields and lower harvesting costs of Brazilian eucalypts, with yields estimated to be 35–40 per cent higher than willow grown in Ontario and harvest costs estimated to be 35–40 per cent lower.

Eucalypt-based energy cropping systems in Australia are also yet to become commercially viable, but have been explored through a range of studies. Stucley et al. (2012) estimated the delivered costs (i.e. plant gate) of mallee eucalypt crops grown in a short rotation coppice (SRC) in Western Australia (WA) to be AU$53–70 (approx. US$40–53) per green tonne, including the costs of growing, harvesting and transporting the biomass, as well as the opportunity cost of the land. Analysis of six bioenergy options for the central west of New South Wales (Baumber, 2012) found that the production of briquettes or wood pellets, and possibly cellulosic ethanol, may be able to pay biomass prices in the range identified by Stucley et al. (Figure 7.3).

The wood pellet and briquette options that appear most viable in Figure 7.3 are likely to be small-scale and subject to price volatility and competition from other sources of biomass. For a larger-scale energy cropping industry to be developed, options such as electricity generation, cellulosic ethanol or renewable diesel are likely to be required, with cellulosic ethanol appearing to be the closest of these three options in Figure 7.3 to meeting the estimated costs of supplying biomass. However, analysis of mallee cropping by Baumber (2012) also found that the production costs for New South Wales were likely to be very different to those estimated by Stucley et al. (2012) for Western Australia. In NSW, a breakeven biomass price of around AU$90–100 rather than AU$53–70 was estimated to be required to cover the opportunity costs of taking the land out of wheat and sheep production.

The biggest factor in the different estimates for NSW and WA is the much lower mallee yields expected for the NSW central west compared to the southwest of WA. A complementary analysis (Baumber et al., 2012) showed that yields were likely to be higher if production was shifted to the higher-rainfall

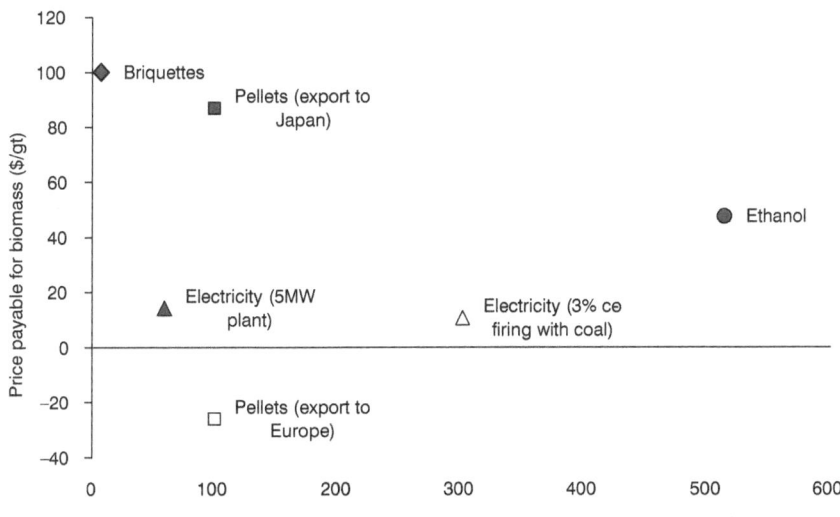

Figure 7.3 Biomass price payable to growers and scale of biomass use for energy cropping modelling for the central west of NSW

Source: Baumber (2012)

Note: Results are for base case assumptions involving processing in the town of Condobolin. Renewable diesel refers to diesel substitutes produced through thermo-chemical conversion of biomass. Prices are shown in Australian dollars per green tonne.

central tablelands of NSW (and based on different species such as Tasmanian blue gum), but that the competing use of intensive sheep-grazing also created a higher opportunity cost.

Valuing the environmental benefits of energy crops

The economic analysis presented so far in this chapter relates primarily to the direct costs of growing biomass for biofuels and electricity and the direct benefits received from the sale of these products. However, as discussed in Chapter 4, energy cropping systems are also capable of providing a range of other benefits, such as carbon sequestration, salinity mitigation, phytoremediation, watershed protection and habitat provision. These outcomes are termed *environmental externalities* due to the fact that their benefits are felt beyond those who are directly undertaking the actions that cause them and they are not captured in traditional markets for goods and services.

It is possible to estimate the value of the environmental externalities provided by energy cropping and other land uses and to design policy measures that reward those who provide them (and create incentives to provide more). Such policy measures are often termed payments for environmental (or ecosystem) services (PES). According to Wunder (2005), the criteria for PES are that the arrangement is voluntary, involves at least one "seller" and one "buyer", and is conditional on the delivery of a well-defined environmental service (or land use activity likely to secure that service). The most common environmental service covered by PES schemes is carbon sequestration, but it is also possible to design measures that provide payments for biodiversity conservation, watershed protection, salinity mitigation or soil protection.

Carbon taxes and emissions trading schemes provide an opportunity to reward landholders for the carbon sequestration services that they may provide. Such schemes employ a "polluter pays" model that requires emitters of greenhouse gases to hold permits covering their emissions. However, they often allow emitters to "offset" their emissions instead by paying landholders for carbon sequestration services. Australia's Carbon Farming Initiative (now part of the Emissions Reduction Fund) is a good example of a scheme that pays landholders for sequestering carbon, including through farm forestry plantings that are subject to periodic harvest.

Stucley et al. (2012) estimated that receiving payments for the carbon sequestration benefits of mallee cropping could improve the economic benefits of mallee cropping by up to AU$2.17 for every green tonne of biomass produced. However, even greater benefits were estimated for other environmental co-benefits, such as reduced waterlogging (up to AU$10.10 gt^{-1}) and protecting biodiversity and public assets from salinity (up to AU$2.80 gt^{-1}). For willow crops grown on contaminated land in Europe, the value of phytoremediation services has been estimated at around €14,000 per hectare over a 20-year period (Lewandowski et al., 2006). Obtaining payments for reducing salinity, protecting biodiversity or removing contaminants is more challenging than payments for carbon sequestration, but there are a number of examples around the world of these kinds of benefits being assigned a monetary value.

As discussed in Chapter 4, Costa Rica is a country that has become well known internationally for its PES model, which has succeeded in directing voluntary payments from private companies (mostly hydroelectric plants) to landholders managing land for not only carbon sequestration, but also for watershed protection, biodiversity conservation and landscape beauty (Porras et al., 2013). Biodiversity offsetting schemes have been implemented in Australia and the US, allowing developers to clear land in return for paying landholders for the protection or restoration of land elsewhere. Land degradation offsetting is another option which, despite being controversial, could direct payments to landholders who reverse degradation on their land (Tal, 2015).

Apart from complex measures involving tradable credits and offsets, there are also a range of simpler schemes that can provide government payments for restorative actions. Such measures may be explicitly linked to a particular environmental outcome or designed to cover a range of outcomes. For example, the per-hectare subsidy paid to energy croppers under the EU's Common Agricultural Policy (CAP) is designed to act as a form of "carbon payment" for the services such crops provide by sequestering carbon and reducing the use of fossil fuels. However, unlike a carbon offset, it is paid on a per-hectare basis rather than per unit of carbon sequestered. Similarly, US Department of Agriculture incentives are paid on a per-hectare basis for the conversion of land to more permanent cover, such as perennial energy crops, for a variety of benefits such as reduced soil erosion, improved water quality, wildlife habitat and enhanced forest and wetland resources (El Kasmioui and Ceulemans, 2012).

A third option for valuing ecosystem services, apart from directing payments from either polluters or governments to landholders, is to factor these values into the renewable energy support systems that already exist. For example, in countries where biofuels are promoted through reductions in excise tax, the level of reduction could be higher for those fuels that can demonstrate a benefit for land. Alternatively, where mandates are used, a system of "double-counting" could be applied to fuels with environmental co-benefits. These options are discussed further in Chapter 8, which reviews the range of policy measures discussed in the book so far, and Chapter 9, which focuses on case studies for Australia and Brazil.

Conclusion

Understanding the economic factors that make particular energy crops viable or unviable is a critical step towards realising the vision of energy crop sustainability. Even for those who do not have a direct economic stake in the successful deployment of energy crops, the environmental and social benefits that these crops may offer are dependent on the successful development of economically viable production models. This requires careful consideration of both the processes involved in growing bioenergy feedstocks, as well as the conversion of such feedstocks to marketable forms of energy, such as transport fuels and electricity.

At present, the most economically viable forms of energy crops are ethanol and biodiesel produced from common agricultural crops, along with electricity generated from woody SRC crops. This may change over time as advances are made around cellulosic biofuels, but progress to date has been slower than expected. Policy measures such as tax breaks and mandates have been critical in achieving cost reductions for first-generation biofuels such as corn ethanol, and they can be expected to play a similarly important role around emerging technologies, helping to bring down costs through learning-by-doing. However, such cost reductions rarely follow a linear path, with years of slow progress often followed by frenzied growth once thresholds for change are crossed.

As the economics change for different forms of bioenergy, there is likely to a shift away from the use of food crops for biofuels and away from electricity being the dominant market for cellulosic crops. However, we can also expect to see surprises, as feedstock producers and energy suppliers are likely to follow the most profitable path laid out by economic conditions, rather than the path laid out by policy-makers. The role played by energy croppers and biofuel processors in developing countries may well present the greatest unknown, as the development of new technologies help to unlock their vast potential for low-cost feedstocks.

One of the great untapped potentials of bioenergy policy around the world remains its inability to properly value the environmental co-benefits that can flow from well-designed and well-executed energy cropping systems, especially those involving perennial trees and grasses. While the replacement of fossil fuels (and the resulting reduction in greenhouse gas emissions) has been a clear target of bioenergy policy in many countries, there is yet to be as strong a recognition of other benefits, such as soil protection, biodiversity conservation, phytoremediation or salinity mitigation. Placing an appropriate economic value on these co-benefits of energy cropping presents an important opportunity in coming years. This could take the form of alternative policies to those currently being used to promote bioenergy, or modifications to existing policies that explicitly preference fuels with environmental benefits. These potential future directions for bioenergy policy are the focus for the final three chapters of the book.

References

Baumber, A. (2012) "Harnessing bioenergy as a driver of revegetation: an analysis of policy options for the New South Wales Central West, Australia", PhD thesis, University of New South Wales, Sydney.

Baumber, A., Rammelt, C., Ampt, P. and Merson, J. (2012) *Bioenergy from Native Agroforestry: Planning for a Regional Industry in the NSW Central Tablelands*, Rural Industries Research and Development Corporation, Canberra.

Brown, T. R. (2015) "A techno-economic review of thermochemical cellulosic biofuel pathways", *Bioresource Technology*, 178: 166–76.

Chen, X. and Khanna, M. (2012) "Explaining the reductions in US corn ethanol processing costs: Testing competing hypotheses", *Energy Policy*, 44: 153–59.

Clean Energy Regulator (2015) *Auction Results: April 2015*, Clean Energy Regulator, http://www.cleanenergyregulator.gov.au/ERF/Published-information/auction-results/auction-results-april-2015 (accessed 22 June 2015).

Commonwealth of Australia (2011) *Securing a Clean Energy Future: The Australian Government's Climate Change Plan*, Canberra, Department of Climate Change and Energy Efficiency: 135.

Department of Energy and Climate Change (2012) "Eligible renewable sources and banding levelS", www.decc.gov.uk/en/content/cms/meeting_energy/renewable_ener/renew_obs/eligibility/eligibility.aspx (accessed 10 March 2012).

Ecofys (2012) *Data and Methodology for Country-Specific Factsheets*, Ecofys, Utrecht.

Ecofys (2013) *How to trigger low carbon technologies by EU targets for 2030?*, Ecofys, Utrecht.

El Kasmioui, O. and Ceulemans, R. (2012) "Financial analysis of the cultivation of poplar and willow for bioenergy", *Biomass and Bioenergy*, 43: 52–64.

Energy Information Administration (2012) *Biofuels Issues and Trends: October 2012*, Energy information Administration, Washington, DC.

Energy Information Administration (2015) "Table 10.3: fuel ethanol overview", www.eia.gov/emeu/mer/pdf/pages/sec10_7.pdf.

Environmental Protection Agency (2010) *Regulation of Fuels and Fuel Additives: Changes to Renewable Fuel Standard Program; Final Rule*, 40 CFR Part 80 [EPA–HQ–

OAR–2005–0161; FRL–9112–3] RIN 2060–A081, United States Environmental Protection Agency, Washington, DC.

Environmental Protection Agency (2013) *EPA Proposes 2014 Renewable Fuel Standards, 2015 Biomass-Based Diesel Volume*, United States Environmental Protection Agency, Washington, DC.

Gerasimchuk, I. and Koh, P. Y. (2013) *The EU Biofuel Policy and Palm Oil: Cutting Subsidies or Cutting Rainforest?*, International Institute for Sustainable Development, Winnipeg, OH.

GSI-IISD (2013) *Addendum to Biofuels – At What Cost? A Review of Costs and Benefits of EU Biofuel Policies*, Global Subsidies Initiative–International Institute for Sustainable Development, Winnipeg, OH.

Hettinga, W. G., Junginger, H. M., Dekker, S. C., Hoogwijk, M., McAloon, A. J. and Hicks, K. B. (2009) "Understanding the reductions in US corn ethanol production costs: an experience curve approach", *Energy Policy*, 37: 190–203.

Langeveld, J. W. A. and Quist-Wessel, P. M. F. (2014) "Biofuel production in Brazil", *In:* Langeveld, J. W. A., Dixon, J. and Keulen, H. v. (eds), *Biofuel Cropping Systems: Carbon, Land and Food*, Routledge, Abingdon, 63–86.

Langeveld, J. W. A., Quist-Wessel, P. M. F. and Croezen, H. (2014) "Biofuel production in the Far East", *In:* Langeveld, J. W. A., Dixon, J. and Keulen, H. v. (eds), *Biofuel Cropping Systems: Carbon, Land and Food*, Routledge, Abingdon, 159–73.

Lewandowski, I., Schmidt, U., Londo, M. and Faaij, A. (2006) "The economic value of the phytoremediation function – assessed by the example of cadmium remediation by willow (*Salix* ssp)", *Agricultural Systems*, 89: 68–89.

Méjean, A. and Hope, C. (2010) "Modelling the costs of energy crops: a case study of US corn and Brazilian sugar cane", *Energy Policy*, 38: 547–61.

Ong, H. C., Mahlia, T. M. I., Masjuki, H. H. and Honnery, D. (2012) "Life cycle cost and sensitivity analysis of palm biodiesel production", *Fuel*, 98: 131–9.

Porras, I., Barton, D. N., ., Miranda, M. and Chacón-Cascante, A. (2013) *Learning from 20 Years of Payments for Ecosystem Services in Costa Rica*, International Institute for Environment and Development, London.

Soto, I., Feto, A. and Keane, J. (2013) *Are Jatropha and Other Biofuels Profitable in Africa?* Jatropha Facts Series, Issue 4, ERA-ARD, www.bioenergyinafrica.net/fileadmin/user_upload/documents/BIA_presentations/4_profitability.pdf.

Stephen, J. D., Mabee, W. E. and Saddler, J. N. (2013) "Lignocellulosic ethanol production from woody biomass: the impact of facility siting on competitiveness", *Energy Policy*, 59: 329–40.

Stucley, C., Schuck, S., Sims, R., Bland, J., Marino, B., Borowitzka, M., Abadi, A., Bartle, J., Giles, R. and Thomas, Q. (2012) *Bioenergy in Australia: Status and Opportunities*, Bioenergy Australia, Killara.

Tal, A. (2015) "The implications of environmental trading mechanisms on a future Zero Net Land Degradation protocol", *Journal of Arid Environments*, 112, Part A: 25–32.

van Eijck, J., Smeets, E. and Faaij, A. (2012) "The economic performance of jatropha, cassava and Eucalyptus production systems for energy in an East African smallholder setting", *GCB Bioenergy*, 4: 828–45.

Wunder, S. (2005) *Payments for Environmental Services: Some Nuts and Bolts*, Center for International Forestry Research (CIFOR), Bogor.

Part IV

Moving forward

Chapter 8

Review of policy options

The preceding chapters have highlighted the variety of environmental, social and economic issues that help to frame our conceptualisations of bioenergy sustainability. Along the way, a range of examples have been provided showing how the development of energy crops can be guided (or misguided) through the use of public policy. This chapter seeks to systematically explore the most relevant policy options for promoting and guiding energy cropping, which are then evaluated for the two national case studies in Chapter 9.

This chapter starts with an overview of the main categories of policy instrument that can be used by policy-makers to promote sustainability. The focus then narrows to the types of measures that can be used to promote sustainability around energy cropping, drawing on the various policy examples cited in the preceding chapters. The discussion of different policy options includes how incentives and disincentives can be delivered most effectively (e.g. using "command-and-control" or "market-based" instruments) and what modifications could be made to common policy options. The final section revisits a question that has arisen in a number of the preceding chapters – is energy cropping a unique sustainability challenge that requires specific policy measures or can it be managed using broader policy measures that apply to variety of land use activities?

Types of policy instruments

There are numerous ways to categorise the various policy instruments that can be used to promote sustainability. Table 8.1 shows two such categorisations, from Stephen Dovers (2005) and David Pannell (2008). Neither of these frameworks are specific to bioenergy. Dovers' framework is designed to promote sustainability in a wide range of contexts and includes fifteen categories covering research and development, regulatory measures, market mechanisms and inaction (which may be the most appropriate choice where the costs of taking action outweigh any benefits). Pannell's framework is designed around sustainable land management and includes only five options: positive incentives, negative incentives, extension, technology change and no action (i.e. informed inaction).

Pannell's *positive incentive* category includes a range of measures designed to promote actions that enhance sustainability, many of which fall under Dovers' market mechanisms category, such as subsidies, tax breaks, mandates and tradable credit schemes. *Negative incentives* are used to discourage actions that threaten sustainability and commonly involve regulatory responses (i.e. Dovers' categories of statute law, common law, conventions, covenants and self-regulation). However, this dichotomy (market-based = positive, regulatory = negative) is far from absolute, as market mechanisms can also feature negative incentives, such as fines, taxes or permits to pollute. Similarly, regulatory measures can be used to reinforce positive actions, such as the incentive that exists to manage land for long-term outcomes when property rights are secure under the law.

Pannell's category of *technology change* is suited to circumstances where no sustainable land use option exists that would be cost-effective for landholders to adopt or for governments to incentivise. This corresponds most closely to Dovers' research and development category, although there is also an important role for market mechanisms such as mandates and subsidies in helping new technologies to achieve commercialisation, as discussed in relation to US corn ethanol in Chapter 7. Where appropriate technologies already exist and financial incentives are not required to drive adoption, *extension* may be the most appropriate solution. Extension-based approaches cut across a number of Dovers' policy classes, including education and training, communication strategies, consultation and community involvement.

There are a number of categories in Dovers' framework that do not have a clear counterpart in that of Pannell. Some of these relate to situations where insufficient knowledge or conflicting viewpoints prevent policy-makers from making a clear choice between positive incentives, negative incentives, technology change, extension or no action. In such cases, efforts to assess impacts, improve communication and information flows, increase community participation and support new research may assist with the policy-making process. In other cases, there may be barriers to the effective delivery of incentives or other policy measures, creating a need for institutional change or the removal of other distorting policies.

Moving forward through this chapter and the next, a hybrid categorisation of policy instruments will be employed, featuring Pannell's five categories, plus an additional two categories that cover the policy classes from Dovers that are not well captured by Pannell's framework. These two additional categories are termed "knowledge and values gathering" (covering situations where further information and perspectives are required in order to make decisions) and "broader institutional reforms" (covering changes to institutions, organisations and other policy arrangements to ensure the smooth operation of selected policy measures).

Table 8.1 Policy instrument categories from Stephen Dovers and David Pannell

Policy instrument class from Dovers (2005)	Corresponding categories from Pannell (2008)	Examples
1. Research and development	• Technology change	• Basic research (general) • Applied research (specific)
2. Creating new or improving existing communication and information flows	• Extension	• Sustainability indicators • State of the environment reporting • Community-based monitoring
3. Education and training	• Extension	• Education may be public, targeted or formal • Training (skills development)
4. Consultative		• Mediation, negotiation
5. Agreements and conventions	• May cover any of the five categories	• Intergovernmental agreements • Conventions, treaties
6. Statute law	• Negative incentives • Positive instruments	• Prohibit or limit damaging practices (negative) • Protect property rights (positive)
7. Common law	• Negative incentives	• Lawsuits for negligence, nuisance
8. Covenants on title	• Negative incentives • Positive incentives	• Prohibit damaging land use options (negative) • Provide stewardship payments (positive)
9. Assessment procedures		• Review of effects • Environmental impact assessment • Life cycle assessment
10. Self-regulation	• Negative incentives • Positive incentives	• Code of practice backed up by disciplinary measures (negative) • Industry standards that facilitate access to new markets (positive)
11. Community involvement	• Extension	• Policy development participation • Community management
12. Market mechanisms	• Negative incentives • Positive incentives • Technology change	• Taxes, charges, fines (negative) • Subsidies (positive) • Tradable quotas (both)
13. Institutional or organisational change		• New or revised settings to enable policy implementation
14. Change other policies		• Remove distorting subsidies or conflicting policies
15. Inaction	• No action	• Informed inaction, with commitment to reconsider over time

Policy instruments for energy crop sustainability

Looking back over the examples cited in the preceding chapters, there are policies relating to energy cropping that fall into each of seven policy categories developed for use in this chapter (negative incentives, positive incentives, extension, technology change, knowledge and values gathering, broader institutional reforms and no action). These examples are summarised in Table 8.2, followed by further discussion of the various ways that policy instruments can be tailored to create the right mix of incentives to promote sustainable energy cropping while restricting unsustainable activities.

Command-and-control versus market-based instruments

The categorisation of policy instruments in Table 8.2 is largely based on whether they seek to encourage or discourage certain actions, or change the range of options available through technology change or knowledge gathering. However, it also important to consider which policy measures can be used to create the desired mix of incentives in the most cost-effective and economically efficient manner. One key debate around cost-effectiveness and efficiency revolves around the use of "command-and-control" versus "market-based" instruments.

Many of the negative incentives listed in Table 8.2 could be considered command-and-control measures, such as the regulatory controls on forest-clearing in Indonesia and Brazil. These measures work by making certain undesirable actions illegal, with penalties such as fines, land seizure or imprisonment to act as deterrents. Measures of this nature provide an important foundation for a sustainable future by placing clear limits on the destruction of the Earth's natural capital for private gain. However, as discussed for Indonesia and Brazil in Chapter 3, there have been significant challenges in relation to monitoring compliance with these restrictions and imposing penalties when laws are breached. In the case of Brazil, a slowdown in deforestation has only been achieved by complementing the regulatory restrictions on forest-clearing with market-based approaches, such as providing incentives for compliance and promoting supply chain agreements among buyers of soybeans and beef (Stickler et al., 2013; Nepstad et al., 2014).

Market-based policy measures differ from command-and-control measures in that they are generally applied in a more flexible manner and seek to use economic incentives to influence behaviour rather than defining actions as legal or illegal. In addition to influencing the behaviour of those who impact upon the natural environment, market-based measures can also be designed to charge the users of natural resources and/or compensate those who provide public benefits such as biodiversity conservation, soil protection or maintenance of water quality. The term *environmental externalities* refers to the costs or benefits related to an action that are not experienced directly by those undertaking the action and are not captured in traditional markets for goods and services. For example, the loss of soil and water protection services when forests are cleared represents a *negative externality*, while certain types of energy cropping involving willow, mallee or switchgrass have the potential to create *positive externalities* by reducing emissions of greenhouse gases, protecting soils, improving water quality and creating habitat for biodiversity.

At their simplest, market-based instruments may take the form of grants or tax breaks designed to encourage a given action, whether related to energy cropping (e.g. the tax breaks for biodiesel provided by Brazil's PNNB programme) or not (e.g. CRP payments in the US). In the case of energy crops, grants or tax breaks may be targeted at various points in the production chain, such as the establishment of plantations, the processing of feedstocks into biofuels or the point at which biofuels enter the transport fuel supply.

Most grants programmes targeted at bioenergy are run by government agencies, but other grants relating to social or environmental outcomes from land use activities may be offered by non-government organisations (NGOs). As discussed in Chapter 4, international NGOs such as the World Wide Fund for Nature (WWF) play an important role in funding restoration projects in developing countries, along with inter-governmental agencies such as the United Nations Development Programme (UNDP).

Negative incentives can also be market-based, including financial disincentives used to discourage particular actions. Examples include carbon pricing schemes that require emitters to hold permits or pay a fixed price (i.e. a "carbon tax") for the right to emit greenhouse gases. In these cases, the price that is placed on the undesirable action may be viewed as a form of compensation to broader society for the public costs (i.e. negative externalities) that these actions create.

The positive or negative incentives offered by market-based instruments need not be purely monetary in order to influence behaviour. In Brazil, access to agricultural credit and insurance has been used to provide an incentive for Amazon landholders to comply with forest protection laws (Butler, 2011). Enhanced security of land tenure can also act as an incentive to protect or restore land in situations where tenure is insecure, with an example being the Sumberjaya pilot programme aimed at watershed protection in Indonesia (OECD, 2010). Conversely, denying access to assistance programmes or making land tenure less secure may act to discourage actions that are seen as undesirable. In these cases, the line between command-and-control and market-based philosophies can become blurred, as such measures may be viewed as either regulatory restrictions backed up by penalties for non-compliance, or economic incentives being applied selectively to increase the attractiveness of one action over another.

Many of the simpler measures shown in Table 8.2 can be implemented by a single government agency to either encourage or discourage a given action. However, as policy measures increase in complexity and become more "market-based", they often involve the cooperation of numerous agencies (potentially across jurisdictions) and create a diverse range of positive and negative incentives simultaneously. The following sections delve into the complexities of a number

Table 8.2 Policy examples cited in the preceding chapters of this book

Policy category	Examples relating to energy cropping (with relevant chapter)
Negative incentives	• Costs imposed on greenhouse gas emissions under cap-and-trade schemes such as the EU Emissions Trading Scheme (Chapter 2) • Indonesian moratorium on forest-clearing (Chapter 3) • Brazilian forest code restricting forest-clearing (Chapter 3) • Certification schemes such as EU RED, SAN, FSC and RSB that provide a barrier to entry to certain markets for products associated with deforestation, high greenhouse gas emissions or high social costs (Chapters 3–6) • Restrictions imposed on the expansion of biofuel production to protect food security in countries such as China (Chapter 5) • Proposed restrictions on biofuel expansion to "idle land" to protect food security and forests (Chapter 5)
Positive incentives	• Biofuel mandates in Brazil, the US, UK and numerous other countries (Chapters 2 and 7) • Preferential treatment of cellulosic biofuels under the EU Renewable Energy Directive (RED) through "double-counting" (Chapter 2) • Tax breaks to promote biofuels such as US ethanol (Chapter 7) or to offset the costs of plantation establishment (Chapter 4) • Certification schemes such as EU RED, SAN, FSC and RSB that provide access to new or higher-value markets (Chapters 3–6) • Grants for land use activities that promote ecological restoration, such as the CRP in the US or BushTender in Victoria, Australia (Chapter 4) • Non-financial incentives for forest protection, such as enhanced security of tenure and access to insurance and credit (Chapters 4 and 6) • Sale of carbon credits from carbon sequestration by mallee eucalypts or other energy crops (Chapters 2 and 4) • Potential sale of biodiversity or ecosystem credits from energy cropping systems (Chapter 4)
Extension	• Need to assist landholders in overcoming barriers to effective cultivation of emerging crops, such as jatropha (Chapter 4) • Extension of knowledge around growing mallee with grass or mulch in between rows to enhance water infiltration (Chapter 4) • Providing seeds and other assistance to increase smallholder participation in biodiesel production in Brazil (Chapter 6)

Technology change	• Research and development into advanced biofuels (Chapter 1) • Double-counting of advanced biofuels under EU RED to promote their development (Chapter 2) • Progressive increase in percentage greenhouse gas saving required from biofuels under US Renewable Fuel Standard (Chapter 2) • Cost reductions brought about through "learning by doing" in supported industries, such as US corn ethanol (Chapter 7)
Knowledge and values gathering	• Greenhouse gas life cycle assessment (LCA) under EU and US biofuel mandates (Chapter 2) • Debates around whether to include indirect land use change in greenhouse gas assessment of biofuels (Chapter 2) • Research into which policy measures have been most effective in slowing deforestation in Brazil (Chapter 3) • Research into environmental benefits and costs of energy crops such as willow, mallee eucalypts, jatropha and switchgrass (Chapter 4) • Debates around the value of "food versus fuel" and research into land availability (Chapter 5) • Social analysis to determine community perspectives and preferences regarding energy crop expansion (Chapter 6)
Broader institutional reforms	• EU decision to endorse other sustainability certification schemes rather than having to assess every biofuel producer directly (Chapter 2) • Enhancing security of tenure to provide incentives for long-term management (Chapter 6) • Enhancing the role of Brazil's state-owned oil company in overcoming barriers to landholder participation in biodiesel production (Chapter 6) • Removal of perverse incentives, such as the incentive for Indonesian palm oil companies to focus on forested areas to avoid land rights negotiations (Chapter 6)
No action	• EU decision not to impose sustainability criteria on solid and gaseous forms of bioenergy due to perceived lower risk (Chapter 2) • Decision by UK Government not to restrict biofuel production to "idle land" as proposed by the Gallagher Review (Chapter 5)

of leading market-based instruments with applicability to the energy and land use sectors. These include auction-based approaches, tradable credit schemes, mandates, feed-in tariffs and certification schemes. However, it is important to remember that the classification of these measures as "market-based" does not mean that they are diametrically opposed to policies based on regulatory restrictions. In many cases, government regulation is the key first step (e.g. imposing caps on greenhouse gas emissions or mandating the use of biofuels), with the power of markets then harnessed to allow these objectives to be met in a more flexible and efficient manner.

Efficiency, cost-effectiveness and auction-based approaches

A key issue surrounding both positive and negative incentives is efficiency. For example, offering the same tax break to all ethanol producers may not be the most efficient way of lowering greenhouse gas emissions, as GHG savings can vary significantly between feedstocks and production processes (as shown in Chapter 2). Similarly, offering the same sized grant to all energy croppers may not be an efficient way to maximise environmental outcomes, as the impact on soil health, water quality and biodiversity can vary enormously between energy cropping systems. This logic also applies to negative incentives such as regulatory restrictions on land clearing or greenhouse gas emissions, as the costs and benefits associated with these activities will vary between different contexts.

Auction-based approaches are one method that can be used to increase the efficiency of incentive-based policy measures. The EU's Emissions Trading Scheme provides an example of this, selling emission permits to polluters via a public auction rather than for a fixed price. This approach avoids the risk that the regulatory authorities might under or over charge for these permits, with the price at auction being determined by the level of demand among polluters for permits and the price at which they would rather reduce their emissions than buy additional permits.

Auction approaches have also been used to improve the cost-effectiveness of grants for environmental benefits. The Conservation Reserve Program (CRP) in the US and the BushTender programme in the Australian state of Victoria cited in Chapter 4 both employ an auction approach. The process involved in these schemes is more accurately described as an *inverse auction* or *reverse auction*. Unlike a traditional auction, where multiple bidders compete to purchase a single item, the process involves multiple providers of restoration services competing for a fixed pool of government funds.

Auctions can help to overcome the risks posed by "information asymmetry", whereby landholders bidding for grants (or polluters bidding for permits) have a better understanding of the true costs of their actions than the government agencies assessing them (OECD, 2010). The OECD (2010) analysed a number of case studies where reverse auctions have been used to distribute environmental grants, include the Tasmanian Forest Conservation Fund in Australia, the Conestoga watershed protection scheme in the USA and the Sumberjaya watershed pilot in Indonesia. They found a strong case for reverse auctions enhancing the cost-effectiveness of grants programmes, including a seven-fold increase in phosphorous reduction per dollar spent in the Conestoga example and a 52 per cent cost-effectiveness gain in the case of the Tasmanian Forest Conservation Fund (compared to allocating grants on a "first-come first-served" basis). The Sumberjaya pilot programme in Sumatra, Indonesia, is notable for the fact that it was NGO-funded (World Agroforestry Centre) and that it involved active revegetation rather than simply the protection of remnant vegetation.

A recent review by the US Department of Agriculture found that the use of auctions to distribute conservation grants can be more effective in terms of reducing costs and maximising environmental benefits than other mechanisms, such as offering a single fixed price to landholders (Hellerstein et al., 2015). However, they also suggested reforms to some elements of the CRP, particularly the use of bid caps, which are designed to prevent landholders making excessive profits. Easing restrictions on how grant money may be used (e.g. for direct costs versus profits) has the potential to make a scheme more attractive to entrepreneurial landholders who can provide cost-effective restoration for a profit, but who would not apply if the scheme only covers a portion of direct costs.

As discussed in Chapter 4, auctions or other grants schemes that attempt to compare projects with diverse outcomes in terms of biodiversity, soils and water quality will inevitably face challenges around how to weigh up these benefits on a common scale. Comparing all projects on a common scale can potentially disadvantage projects with unique outcomes that cannot easily be compared to other projects. One solution to this problem under the CRP is to allow a relatively small number of sites with unique characteristics to join through a non-competitive continuous sign-up process, which is aimed at protecting land with the greatest conservation value, regardless of whether such sites would rank highest in terms of cost-effectiveness under the auction-based general sign-up process.

Other challenges with auction approaches include the risk that offering payments will deter voluntary action that would have taken place without any payment (known as "crowding-out") and the risk that payments will be made for projects that would have happened anyway (Hellerstein et al., 2015). This latter problem is often referred to as a failure to ensure the "additionality" of projects received funding and can reduce the cost-effectiveness of an auction scheme. In the case of conservation grants schemes, it is also essential to have measures in place to ensure that the benefits predicted at the time the grant was issued are actually realised, including some form of monitoring and verification processes (i.e. the "metrics" discussed in Chapter 4) and mechanisms for rescinding payments in cases of non-compliance.

Tradable credit schemes

While auction approaches have the potential to enhance the efficiency of policy incentive schemes, they still employ the model of a single buyer or seller (generally a government agency) who is either paying for public benefits or charging for public costs. However, policy measures can also be designed in such a way as to incorporate multiple buyers and sellers and both positive and negative incentives. In theory, approaches that involve multiple buyers and sellers are more "market-based" and should result in more efficient allocation of resources by enhancing competition. However, they can also provide another key benefit for those planning actions such as energy cropping that may result in public benefits for biodiversity or ecosystem health. Schemes that incorporate private buyers of ecosystem services can provide an alternative source of funding that sidesteps the traditional reliance on grants from government agencies or environmental NGOs. Harnessing the capacity of businesses and wealthy individuals to pay for the services they derive from managed ecosystems offers the potential to greatly expand the pool of funding available for restoration activities.

The role that tradable credit schemes can play in incentivising carbon sequestration and biodiversity conservation was discussed in Chapters 4 and 7. Carbon trading schemes generally operate by requiring emitters of greenhouse gases to hold permits covering their emissions. As discussed in Chapter 2, this can provide an advantage for electricity generators using bioenergy rather than fossil fuels such as coal. However, such schemes can also incentivise carbon sequestration if emitters are allowed to use sequestration credits to "offset" their emissions. Box 8.1 highlights the way that Australia's approach to carbon offsetting has shifted over recent years. Ironically, rather than following a progression from simpler to more complex schemes over time, the trend in Australia has been the opposite due to political considerations.

In addition to national schemes, such as the example discussed in Box 8.1, the United Nations Framework Convention on Climate Change (UNFCCC) also provides for international carbon trading. These measures were discussed in Chapters 3 and 4, including the potential for investment money to flow from developed to developing countries for reforestation and afforestation projects under the Clean Development Mechanism (CDM). However, the effectiveness of international carbon trading to protect or enhance carbon sinks has been the subject of much debate. Thomas et al. (2010) reported that, out of more than 1600 CDM projects created by 2010, only four were for reforestation or afforestation. They argue that CDM reforms are required to provide greater flexibility, simpler methodological and documentation procedures and a switch in focus from adjudicating to facilitating CDM reforestation projects. Conversely, the REDD+ arrangements have been criticised for allowing too many loopholes, such as the classification of large areas of regrowth forest as "degraded" under the Indonesia–Norway deal, allowing clearing to continue in these areas (Edwards and Laurance, 2011).

Biodiversity and land degradation offsets were discussed in Chapters 4 and 7 as possible ways of incentivising energy cropping systems that contribute to restoration outcomes. One key challenge for biodiversity offsets is ensuring equivalence between the damaging activity that a developer wishes to undertake and the restorative one that is being used as an offset. It is easier to make a case for equivalence in relation to carbon trading, as the Earth's atmosphere is an interconnected global commons and the locations at which CO_2 is added or removed is not particularly important. However, this is not the case for biodiversity outcomes, which are very much dependent on the location at which habitat restoration occurs.

Biodiversity offsetting schemes may attempt to ensure equivalence through a complex set of rules. For example, under the BioBanking scheme in New South Wales (NSW), Australia, developers wishing to destroy habitat receive BioBanking statements that detail not only the number of credits that must be surrendered to offset the habitat destruction, but also the type of credit required (ecosystem or species credits) and the vegetation types in which those credits can be generated. Even despite such rules, many ecologists are sceptical of the capacity of biodiversity offset schemes to ensure equivalence between the areas cleared and those restored (e.g. Gibbons and Lindenmayer, 2007; Burgin, 2008). For these reasons, biodiversity offsetting is unlikely to be a major driver for the establishment of energy cropping systems. Offsets aimed at very specific landscape benefits, such as protection against erosion or salinity from the establishment of mallee plantings in Australia, may be a more viable option for energy cropping, but such a scheme has yet to be demonstrated in practice.

Box 8.1 **The evolution (or regression) of market-based instruments for carbon in Australia**

In the lead-up to 2007 federal election, a bipartisan political consensus emerged that Australia should employ a market-based cap-and-trade approach to reducing greenhouse gas emissions in line with its commitments under the Kyoto Protocol. This scheme, which came to be known as the Carbon Pollution Reduction Scheme (CPRS), would have placed emissions caps on large emitters but allowed trading between them, such that those with excess emission permits could sell them to those wishing to increase emissions. Alternatively, emitters could offset their emissions by purchasing offsets from reforestation projects.

The CPRS was progressed by the newly elected Labor Government through a 2008 White Paper and 2009 negotiations with the opposition Liberal/National coalition, almost making it through parliament before the coalition switched leaders to the anti-CPRS Tony Abbott. Further progress was delayed until after the 2010 election, when a new parliamentary

Continued...

Box 8.1 continued

balance allowed Labor to negotiate a revised carbon pricing model. Unlike the CPRS, which involved placing caps on emitters but letting the price of permits "float" according to demand, the revised model placed no caps on emitters but instead required them to pay a fixed price for permits to the Government (i.e. a "carbon tax"). This price was set initially at \$23 per tonne of carbon dioxide equivalent (tCO_2-e) and was set to rise to \$25.40 per tCO_2-e within three years before transitioning to a floating price (Commonwealth of Australia, 2011). However, this transition to a floating price was never realised, as the scheme was scrapped after the Abbott-led coalition won the 2013 election on a platform of "scrapping the carbon tax".

In terms of market-based instruments, the change from the CPRS to the carbon tax represented a simplification from a multiple buyer/multiple seller model to one in which there were multiple buyers of permits but only one seller (the government) and the price was fixed. However, despite the lack of a competitive market for emissions permits, a competitive market for offsets was created to complement the carbon tax. Under this arrangement, multiple providers of offsets were able to sell to multiple emitters wishing to reduce their carbon tax liability for whatever price the two parties agreed on (with the carbon tax acting as the effective maximum price for offsets). Reforestation and revegetation projects were able to earn offset credits under the Carbon Farming Initiative (CFI), which recognises not only permanent environmental plantings, but also activities that involve future harvesting, such as farm forestry and mallee plantings.

The abolition of the carbon tax in 2014 represented a retreat from placing either caps or taxes on emitters. However, it did not result in the total abandonment of market-based approaches, as the newly created Emissions Reduction Fund (ERF) employed a reverse auction approach to distribute Government funds to providers of emission reductions or sequestration. Importantly (from a restoration perspective), the ERF incorporates the key elements of the Carbon Farming Initiative, allowing reforestation and regeneration projects to be eligible for ERF payments. Indeed, in the initial ERF auction in April 2015, sequestration projects represented around 60 per cent of the 47 million tonnes of abatement purchased by the Australian Government (Clean Energy Regulator, 2015).

Both sides of politics in Australia have argued that their preferred model is the most efficient option for reducing greenhouse emissions at the lowest cost. While the transition from cap-and-trade to carbon tax to reverse auction may not be what most advocates of market-based instruments would anticipate or recommend, a commitment to some form of market-based approach has been an enduring element of Australia's climate change policy in the period 2007–15.

Mandates and feed-in tariffs

As with tradable credit schemes, mandates also seek to combine positive and negative incentives though a market-based approach. Mandates operate by governments dictating to energy suppliers that a set amount of their energy supply must come from eligible renewable sources, which may include various forms of bioenergy. Penalties for failing to meet the mandate provide the negative incentive, while the positive incentive is provided by the demand from energy suppliers to obtain eligible sources of energy. Assuming there are multiple buyers and multiple sellers, a competitive market in eligible fuels should result in no one party being able to dictate prices. In many cases, energy suppliers do not have to deliver the actual energy, but rather are required to surrender a set number of certificates from those who have supplied the eligible energy. This can enhance the efficiency of the market, as certificates are much easier to trade than actual quantities of energy.

Mandates may be used for liquid biofuels, electricity or even the supply of renewable gas or industrial heat. In its 2013 global status report, the renewable energy policy network REN21 lists 30 countries with biofuel mandates at the national or state/provincial level (REN21, 2013). Brazil's pioneering PROALCOOL programme is one of the best-known examples of a biofuel mandate, requiring fuel suppliers to add ethanol to gasoline from the 1970s onwards in order to reduce oil imports and promote economic development in sugar-growing regions (Dufey et al., 2007). Other notable examples include the US Renewable Fuel Standard (RFS), the UK Renewable Transport Fuel Obligation (RTFO) and the Biofuel Act in the Australian state of New South Wales (NSW). The UK RTFO sits beneath the EU's Renewable Energy Directive (RED), with the RED setting the target and the RTFO assigning responsibility to particular fuel suppliers to meet it.

Mandates for renewable electricity are generally broader than those for liquid biofuels, covering multiple forms of renewable energy. For example, schemes such as the UK Renewables Obligation (RO) or Australia's Renewable Energy Target (RET) list energy crops as one eligible source alongside more common forms of renewable generation such as wind, solar photovoltaics and hydropower.

Mandates for both biofuels and electricity have the potential to incorporate differentiated support for different energy sources, also known as "banding". An example of banding for liquid biofuels is the "double-counting" of biofuels from wastes or cellulosic feedstocks under the EU Renewable Energy Directive. With regards to electricity, the UK's Renewables Obligation was cited in Chapter 7 as an example of a mandate scheme that employs a system of banding that provides greater support for energy crops than for more established technologies. As with the double-counting for advanced biofuels under the EU RED, this approach can accelerate the development of new technologies and preference energy sources with a higher level of public benefit.

While banding approaches may help to promote energy crops with environmental co-benefits, a cautionary tale is provided by Australia's use of a form of banding from 2009 to 2010. In this case, small-scale sources such as residential solar photovoltaics were awarded five certificates per MWh rather than just one, due to the fact that they were unable to compete with cheaper large-scale sources such as wind and landfill gas (Department of Climate Change, 2009). While this provided additional incentives for small-scale generation, it also had the effect of flooding the market with "phantom" certificates that did not represent actual generation (Buckman and Diesendorf, 2010). The resulting drop in the price for certificates led to a reduced incentive to invest in other forms of renewable energy such as wind and bioenergy. This problem was eventually addressed through the creation of two separate schemes covering small-scale and large-scale renewables in 2011.

Feed-in-tariffs (FiTs) are an alternative to electricity mandates for promoting renewable generation. Rather than fixing the amount of renewable generation that an energy company must procure, they instead fix the price that such companies must pay for eligible generation. FiTs have been most prominent in increasing uptake of rooftop solar photovoltaic systems in a number of countries (Cory et al., 2009), but they have also been used to promote bioenergy generation in countries such as Germany. The primary advantages of FiTs are that they offer greater price certainty for generators and can be tailored to offer a differing level of support to different technologies. Table 8.3 shows how this differentiation was applied to biomass energy in Germany, with a different tariff paid depending on the size of the generator and the feedstocks used, reflecting the fact that some bioenergy sources have higher costs and some have associated benefits (e.g. landscape preservation or reduced air pollution).

Table 8.3 Example of differentiated feed-in tariffs for bioenergy in Germany

Size tranches		Compensation (€ cents/kWh)
≤ 150 kW		11.67
> 150 kW ≤ 500 kW		9.18
> 500 kW ≤ 5 MW		8.25
> 5 MW ≤ 20 MW		7.79
Bonus Compensation for using:	energy crops	4.0–7.0
	manure	1.0–4.0
	biomass from land managed for landscape preservation	2.0
	innovative technology	1.0–2.0
	combined heat and power	3.0
	fewer emissions (clean air)	1.0

Source: Schuck (2010)

Incentives to combine conservation and commercial production

One mechanism discussed in a number of the preceding chapters for giving preference to products that provide associated environmental benefits is certification against a sustainability standard. The three standards used for comparison in this book have been those of the Forest Stewardship Council (FSC), which is prominent in the forestry sector, the Rainforest Alliance, which utilises the standards of the Sustainable Agriculture Network (SAN), and the Roundtable on Sustainable Biomaterials (RSB), which has developed a set of standards for use in the biofuel sector.

The FSC, SAN and RSB standards are all examples of voluntary standards that rely on demand for "sustainable" products among consumers or other participants in the supply chain. In many cases, businesses may choose to consume certified products to enhance their reputation and demonstrate their commitment to corporate social responsibility. However, the RSB standard differs from the other two in that it has been approved (in a modified form) by the EU to demonstrate compliance with the sustainability criteria of its Renewable Energy Directive (RED). This demonstrates the way in which a mandate scheme can work in conjunction with a certification scheme – and how governments can "outsource" the task of assessing individual energy providers to ensure the sustainability of their fuels.

The idea of using integrated policy instruments to simultaneously promote and restrict different forms of bioenergy has the potential to be taken much further. Table 8.4 lists a number of different policy instruments that have been discussed in the preceding sections, along with examples of where they have been used and ways in which they could be modified to incorporate a preference for feedstock production systems that offer associated environmental benefits.

The approach suggested in Table 8.4 goes further than the example of the RSB (and other) standards being used to restrict the eligibility of biofuels under the EU RED. As discussed in Chapter 4, the EU RED sustainability criteria, along with those of the RSB, FSC and SAN, focus mostly on maintenance rather than enhancement. The examples in Table 8.4 go beyond simply maintaining ecosystem functions by preventing deforestation and degradation and focus instead on actively enhancing such functions.

As shown in Table 8.4, it may be possible to modify incentive schemes involving tax breaks, mandates or feed-in tariffs to preferentially support energy crops that enhance ecosystem services. However, as discussed in Chapter 4, this would require metrics that could effectively predict and monitor the outcomes of eligible energy cropping systems. It is possible that standards such as those of the RSB could be used in conjunction with these modified incentive schemes to promote ecosystem enhancement. However, this may present a conflict between the aim of such standards to promote a level of "best practice" that is achievable by all or most producers, and the aim of rewarding those land managers who go

Table 8.4 Bioenergy support measures with the potential to promote the enhancement of ecosystem functions and services

Policy option	Example	Potential modifications to promote ecosystem services
Tax breaks for biofuel producers	Brazil's biodiesel support scheme, which offers greater tax breaks for "social fuels" that come from small family farmers	While Brazil's scheme seeks to deliver a social benefit, a similar model could be used to preference production systems that enhance ecosystem services
Mandates requiring the use or supply of biofuels	EU mandate requiring fuel suppliers to supply biofuels, which provides greater support to fuels from non-food cellulosic crops through a system of "double-counting"	A similar model of multiple-counting could be used to preference energy cropping systems that enhance ecosystem services
Mandates requiring the supply of renewable electricity (including bioenergy)	UK Renewables Obligation, which includes "banding" that provides higher levels of support for certain options (e.g. energy crops)	Similar to multiple-counting for biofuel mandates, the level of support for biomass crops for electricity could be based on the ecosystem services provided
Feed-in tariffs requiring electricity companies to pay a fixed price for bioelectricity	German feed-in tariffs, which incorporating a bonus for biomass from land managed for landscape preservation	Higher feed-in tariffs could be applied to biomass crops that enhance ecosystem services

beyond best practice and provide positive externalities in relation to biodiversity, soil protection or water quality. This relates to the question raised in Chapter 4 of what we should expect from energy cropping systems, which is revisited in the following section.

Is bioenergy unique when it comes to sustainability policy?

One final question that requires consideration before delving into the case studies in Chapter 9 is whether or not bioenergy is fundamentally different to other land use activities and requires a unique approach when it comes to sustainability. Energy cropping is governed by many of the same policy measures that govern other types of land use, such as land-clearing laws and regulations protecting workers' rights and property rights. However, there are some interesting examples where differing expectations and values are apparent in the policy treatment applied to bioenergy compared to other land uses.

In the preceding sections, examples were given of bioenergy-specific incentives such as tax breaks, mandates and feed-in tariffs, which seek to capitalise on the potential of bioenergy to help mitigate global warming, provide an alternative to imported oil or restore degraded land. In other cases, bioenergy-specific regulations have been used to mitigate risks that are seen to be uniquely posed by certain forms of bioenergy.

In Australia, a particular issue for bioenergy has been the use of native forest residues for electricity generation. Groups such as The Wilderness Society have campaigned against the use of these "wastes" for bioenergy on the grounds that this could increase the overall volume of material taken out of the forest, expand harvesting into old-growth areas and support harvesting that would otherwise be uneconomic, resulting in additional greenhouse gas emissions and loss of biodiversity. In response to these concerns, regulatory restrictions were introduced in 2011 to prevent *any biomass* sourced from a native forest from being counted towards Australia's Renewable Energy Target (RET). This created a situation in which it was legal to harvest native forests for construction timber, paper products or woodchip exports under federally approved "ecologically sustainable forest management" plans, but using the biomass harvested under these plans for electricity generation was considered unsustainable.

Lobbying by groups such as the Australian Forest Products Association led to the RET legislation being amended again in June 2015 to make native forest biomass eligible once more, provided that energy is not the primary purpose of the harvest. The key arguments made by proponents of using native forest biomass for electricity are that similar practices occur in many other countries, the low price of biomass for energy is unlikely to increase the overall level of forest harvesting and the forests in question already have restrictions on both the area of forest and amount of forest material that is available for harvest (Rutovitz and Passey, 2004).

A similar approach can be seen in some of the suggested responses to the "food versus fuel" issue discussed in Chapter 5. One response suggested by NGOs, academics and activists has been to restrict biofuel production to land that is not currently being used for food production, or "idle land" as it was termed in the 2008 Gallagher Review. The common thread between these "idle land" proposals and the Australian forests example is the premise that using land for one purpose (e.g. food) may be sustainable, while using the same land in the same manner to produce the same type of biomass is not sustainable if it is for a different purpose (e.g. transport fuel).

A further example of bioenergy being singled out as a unique threat to sustainability can be found in relation to palm oil. Friends of the Earth UK argued in their 2006 position statement that makers of food and household products containing palm oil should ensure that it is "sustainable" palm oil that has been certified by bodies such as the Roundtable for Sustainable Palm Oil (RSPO). However, in the case of bioenergy, they stated that "since the potential demand for palm oil as a biofuel or for biomass energy is so large and given the weak governance in Indonesia and its destructive policies regarding plantation development, Friends of the Earth does not support the use in the UK of palm oil as a biofuel or for use as biomass for electricity production" (Friends of the Earth UK, 2006, p. 4). Again, there is a differentiation between two sets of products produced from the same land in the same manner, with one set of products (food and household products) being seen as potentially sustainable if certified, while another set of products (biofuels and electricity) are seen as incapable of ever being produced in a sustainable manner.

A common thread in the three examples above is a concern about the potential scale of bioenergy use. In cases where bioenergy threatens to dramatically increase the scale of land or resource use in a way that no other product does, it may indeed be appropriate to create bioenergy-specific regulations to prevent this runaway growth. However, this book has highlighted a number of examples where the size of the threat that bioenergy is seen to pose is disproportionate to its actual share of land or resource use. For example, while the amount of palm oil imported into the EU for bioenergy has shown rapid growth, its overall level of use for energy remains much lower than for food and cosmetics (Gerasimchuk and Koh, 2013). Similarly, while biofuels have been singled out as the leading cause of large-scale land transactions by the International Land Coalition (Anseeuw et al., 2012), the ILC's own Land Matrix shows that most transactions are in fact for food crops (Chapter 6). Also, the analysis by Langeveld et al. (2013) cited in Chapter 5 showed how activities such as urbanisation, mining and infrastructure development can play a greater role in the conversion of agricultural land than biofuels.

Apart from its rate of growth, two other key reasons that bioenergy may attract a disproportionate amount of attention from certain stakeholders are incumbency and perceived value. Bioenergy is often seen as a new option for both energy provision and land use – and this is at least partly true of modern bioenergy in the form of electricity or transport fuel (but certainly not for traditional use of biomass for cooking and heating). At any rate, the perceived newness of bioenergy can result in it being held up to a higher level of scrutiny than incumbent energy sources and land uses. Although there is widespread recognition of the impacts that can result from dominant forms of energy such as coal and oil, as well as from dominant land uses such as agriculture and forestry, it is much easier to impose restrictive measures on an industry that is not yet widespread than on one that involves a vast array of well-established vested interests.

A perception that bioenergy has a low societal value relative to other products may be another factor leading to it being targeted for restrictive measures. This is most apparent in the "food vs fuel" debate discussed in Chapter 5, including proposals to ban further expansion of energy crops or restrict them to "idle land". The question of which land use best meets human needs is indeed a valid sustainability consideration, with the Brundtland definition of sustainable development cited in Chapter 1 making it clear that we must consider how different actions help to meet the needs of present and future generations.

However, while food and bioenergy are far from the only options for any given piece of land, debates around "food versus forestry", "food versus urbanisation" and "food versus conservation" have not attracted the same level of concern as "food versus fuel".

Several of the chapters of this book have highlighted arguments against the idea that bioenergy is always of lower societal value than food or other products. In Chapter 1, UN Energy was quoted as saying that the notion of food versus fuel is "overly simplistic and fails to reflect the full complexity of factors that determine food security at any given place and time" (UN Energy, 2007, p. 31). In Chapter 4, the concepts of multifunctionality and conservation through sustainable use were cited, along with examples of bioenergy-based production systems that actually enhance food security over the long-term. In Chapter 5, the nuanced approach of the Roundtable on Sustainable Biofuels (RSB) was described, under which biofuel producers must enhance local food security when operating in food insecure areas, but biofuel production is not considered automatically to be of lesser value than food production.

Notwithstanding these arguments against imposing land use restrictions exclusively on bioenergy products, there may be circumstances in which it is sensible to vary the way that policy measures are applied to energy crops compared with other crops. These include cases where energy cropping has unique impacts that are not relevant to other crops (e.g. an ability to mitigate global warming by replacing fossil fuels), where the potential scale or rate of expansion of energy cropping dwarfs other products, or where the nature of the stakeholders in the marketplace and/or supply chain are fundamentally different to other products. An example of this latter occurrence was given in Chapter 6, where it was argued that unique biofuel standards are justified by the major role that governments play in driving biofuel demand compared to food, feed or fibre products. Careful consideration is given to whether these circumstances exist for the case studies presented in Chapter 9.

Conclusion

This chapter has highlighted a range of policy measures that may be used to promote, restrict and enhance understanding around a range of land use activities, including energy cropping. These include measures that could be described as "command-and-control" as well as those that are more "market-based", harnessing the power of competitive markets to create incentives and promote cost-effective solutions. As shown through the examples cited, there is no shortage of options for policy-makers to draw on when designing a policy environment in which sustainable energy crops will thrive while unsustainable energy crops are restrained. The challenge is working out which of these options are likely to be effective in specific contexts.

The following chapter attempts to identify policy options that could be adopted and adapted in specific contexts by focusing on two national case studies covering Australia and Brazil. In order to bring together the various issues discussed in this chapter and develop policy options for the two case studies, a policy development framework for sustainable energy cropping based on adaptive management and adaptive governance is presented. By undertaking case study analysis of this nature, we can learn how incentives and disincentives, command-and-control and market-based principles, and energy crops and other land uses can be effectively integrated into a coherent approach to sustainability policy.

References

Anseeuw, W., Wily, L. A., Cotula, L. and Taylor, M. (2012) *Land Rights and the Rush for Land: Findings of the Global Commercial Pressures on Land Research Project*, International Land Coalition, Rome.

Buckman, G. and Diesendorf, M. (2010) "Design limitations in Australian renewable electricity policies", *Energy Policy*, 38: 3365–76 (addendum, 38:7539–40).

Burgin, S. (2008) "BioBanking: an environmental scientist's view of the role of biodiversity banking offsets in conservation", *Biodiversity and Conservation*, 17: 807–16.

Butler, R. (2011) "Could palm oil help save the Amazon?", http://news.mongabay.com/2011/0614-amazon_palm_oil.html (accessed 14 April 2015).

Cory, K., Couture, T. and Kreycik, C. (2009) *Feed-in Tariff Policy: Design, Implementation, and RPS Policy Interactions*, National Renewable Energy Laboratory, Golden, CO.

Department of Climate Change (2009) "Why do we need RET?", www.climatechange.gov.au/government/initiatives/renewable-target/need-ret.aspx (accessed 11 February 2010).

Dovers, S. (2005) *Environment and Sustainability Policy*, Federation Press, Sydney.Dufey, A., Vermeulen, S. and Vorley, B. (2007) *Biofuels: Strategic Choices for Commodity Dependent Developing Countries*, Common Fund for Commodities, Amsterdam.

Edwards, D. P. and Laurance, W. F. (2011) "Loophole in forest plan for Indonesia", *Nature*, 477: 33.

Friends of the Earth UK (2006) "The use of palm oil for biofuel and as biomass for energy: Friends of the Earth's position", www.foe.co.uk/resource/briefings/palm_oil_biofuel_position.pdf (accessed 21 August 2008).

Gerasimchuk, I. and Koh, P. Y. (2013) *The EU Biofuel Policy and Palm Oil: Cutting Subsidies or Cutting Rainforest?*, International Institute for Sustainable Development, Winnipeg, OH.

Gibbons, P. and Lindenmayer, D. B. (2007) "Offsets for land clearing: No net loss or the tail wagging the dog?", *Ecological Management and Restoration*, 8, 26-31.

Hellerstein, D., Higgins, N. and Roberts, M. (2015) *Options for Improving Conservation Programs: Insights From Auction Theory and Economic Experiments*, Economic Research Service, United States Department of Agriculture, Washington, DC.

Langeveld, J. W. A., Dixon, J., Keulen, H. v. and Quist-Wessel, P. M. F. (2013) *Analysing the Effect of Biofuel Expansion on Land Use in Major Producing Countries: Evidence of Increased Multiple Cropping*, Biomass Research, Wageningen.

Nepstad, D., McGrath, D., Stickler, C., Alencar, A., Azevedo, A., Swette, B., Bezerra, T., DiGiano, M., Shimada, J., Ronaldo Seroa da Motta, Armijo, E., Castello, L., Brando, P., Hansen, M. C., McGrath-Horn, M., Carvalho, O. and Hess, L. (2014) "Slowing

Amazon deforestation through public policy and interventions in beef and soy supply chains", *Science*, 6 June: 1118–23.

OECD (2010) *Paying for Biodiversity: Enhancing the Cost-Effectiveness of Payments for Ecosystem Services*, Organisation for Economic Co-operation and Development, Paris.

Pannell, D. J. (2008) "Public benefits, private benefits, and policy mechanism choice for land-use change for environmental benefits", *Land Economics*, 84: 225–40.

REN21 (2013) *Renewables 2013: Global Status Report*, REN21, Paris.

Rutovitz, J. and Passey, R. (2004) *NSW Bioenergy Handbook*, NSW Government, Sydney.

Schuck, S. M. (2010) *Bioenergy Industry: Report June 2010*, Clean Energy Council, Melbourne.

Stickler, C. M., Nepstad, D. C., Azevedo, A. A. and McGrath, D. G. (2013) "Defending public interests in private lands: compliance, costs and potential environmental consequences of the Brazilian Forest Code in Mato Grosso", *Philosophical Transactions of the Royal Society B*, 368: 1616. http://rstb.royalsocietypublishing.org/content/368/1619/20120160 (accessed 5 February 2016).

Chapter 9

Case studies

Australia and Brazil

The wide range of environmental, social and economic issues presented in this book underscore just how complex the question of energy crop sustainability can seem at times. While energy cropping can produce measurable benefits in terms of climate change mitigation, ecosystem health and socio-economic development, it can also lead to unsustainable impacts such as deforestation, degradation and dispossession. Furthermore, many of these impacts are surrounded by a high degree of uncertainty and differing stakeholder values. This can often make it difficult to see a clear pathway forward that maximises the potential benefits of energy cropping while protecting against risks of serious environmental and social damage.

The aim of this chapter is to cut through the complexity surrounding energy crop sustainability to identify feasible pathways forward for two case study nations: Australia and Brazil. These two countries have been chosen because they represent contrasting positions on some of the key aspects of energy cropping. Australia is a developed nation and a member of the Organisation for Economic Co-operation and Development (OECD). In contrast, Brazil is generally classed as an emerging or newly industrialised economy and is a member of the BRICS group (Brazil, Russia, India, China, South Africa). While Brazil has been a global pioneer of energy cropping though its sugarcane ethanol industry, energy cropping remains a minor land use activity in Australia.

In other respects the two countries are similar – although still different enough to make for an interesting comparison. Both are large countries that have significant potential to capitalise on expanding global demand for bioenergy, particularly if cellulosic biofuels become commercially viable. However, Brazil is much closer to the key US and EU markets than Australia and has a track record of exporting biofuels that Australia lacks. Both countries feature a mix of temperate and tropical zones, although Brazil has seen a greater focus on tropical and semi-tropical energy crops compared to Australia's focus on tree crops for low-rainfall temperate regions. Both countries have generated controversy around land-clearing and deforestation, although the level of international concern has been much greater in relation to Brazil. Both countries have also identified energy crops as a potential driver of ecological restoration, such as through the Oil Mallee Project in Western Australia.

The two case studies are reviewed side-by-side throughout this chapter, using an abbreviated version of a sustainability policy framework developed by Stephen Dovers (Figure 9.1). Dovers' framework has four stages: problem-framing, policy-framing, policy implementation and policy monitoring and evaluation (Dovers, 2005). These stages are not strictly linear and the framework features an overarching set of general elements that must be considered as each stage is revisited in an adaptive manner. This chapter lies primarily within stage 4 (policy evaluation), but involves key elements of the earlier stages such as analysing the existing policy environment (stage 1), identifying goals (stage 2) and selecting new policy options or modifications (stage 3). Consideration is also given to how the recommended policy options could be rolled out and what monitoring and review mechanisms are required to ensure that policies are applied in an adaptive fashion.

The main focus of this chapter is on providing advice for policy-makers. However, it necessarily considers actions that could be undertaken by other stakeholders such as landholders, processors, energy suppliers, consumers, investors and non-government organisations (NGOs) to generate the knowledge, technologies, resources and experience necessary for the development of sustainable energy cropping systems. The creation of a suitable policy environment is a critical element in enabling these stakeholders to perform the various roles needed within a sustainable energy cropping industry.

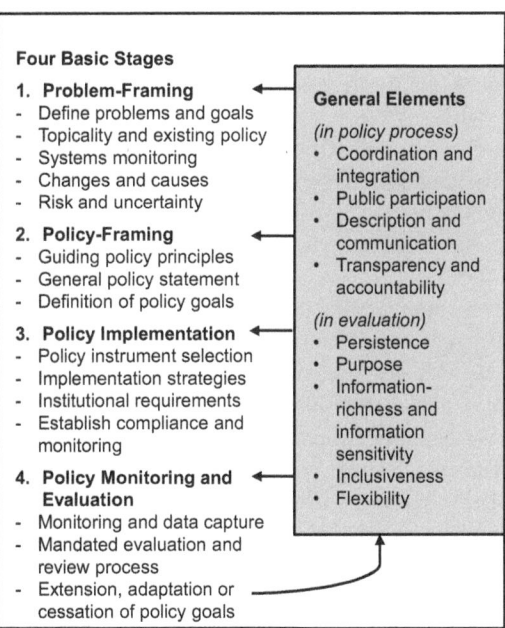

Figure 9.1 Sustainability policy framework from Dovers (2005)

Existing policy conditions in Australia and Brazil

While many of the sustainability issues surrounding energy cropping are common between Australia and Brazil, the social, economic and policy context differs significantly between these two case studies. Similarly, the goals underpinning the development and regulation of energy cropping vary between the two countries as well as across different geographic scales and stakeholder groups within them.

Table 9.1 summarises the main policy measures that influence energy cropping at present in Australia and Brazil. Policies are broken down into the policy categories presented in Chapter 8 (negative incentives, positive incentives, extension, technology change, knowledge and values gathering and broader institutional arrangements). The seventh category (no action) is excluded due to the challenges involved in differentiating between inaction that results from careful analysis and inaction that results from a lack of consideration or political reasons. The list is not intended to be exhaustive, but particularly focuses on those measures that could be modified or adapted to influence the development of energy cropping. Most of the policy measures have been introduced in preceding chapters.

It is worthwhile starting with the final category in Table 9.1, broader institutional arrangements, as these provide the frameworks under which other policy measures sit. Both Australia and Brazil have a federal governance structure, with legislative powers divided between state governments and the federal government (officially the Commonwealth in Australia and the União or Union in Brazil). International conventions also play an important role in policy-setting around energy crops, particularly the United Nations Framework Convention on Climate Change (UNFCCC) and the Kyoto Protocol that sits within it.

Australia is an Annex I (i.e. industrialised) country under the Kyoto Protocol and is thus bound by quantitative emissions targets. Initial targets were set for the period 2008–12, with a subsequent commitment for 2020 made under the 2012 Doha amendment. In contrast, Brazil is a non-Annex I country and is not bound to any quantitative emissions targets. However, it has been an active participant in the Clean Development Mechanism (CDM), under which emissions reductions achieved in non-Annex I countries can be traded to Annex I countries to offset their emissions.

In Australia, the main policy incentives driving bioenergy development are the Renewable Energy Target (RET) and federal rebates on fuel excise for ethanol and biodiesel. The RET is a mandate placed on electricity companies to provide a certain amount of electricity from eligible renewable sources. Liable electricity companies meet their obligations by surrendering a set number of renewable generation certificates. These certificates are traded between renewable generators and liable parties, with the market price varying depending on the prevailing levels of supply (i.e. the amount of renewable generation installed) and demand (i.e. the size of the target, which increases each year to 2020).

"Energy crops" is a specific eligible source category under the RET, but has played a negligible role to date. However, bioenergy produced from co-products of cropping and plantation systems have played a more significant role, with "bagasse" (from sugarcane) and "wood waste" contributing around 7 and 2 per cent of all Renewable Energy Certificates by 2012 respectively (Office of the Renewable Energy Regulator, 2012). The definition of "wood waste" was amended in 2015 to allow some native forest material to be used (after being excluded from 2011 to 2015). This change has been controversial (as discussed in Chapter 8), but has limited relevance to energy cropping.

The RET has been amended several times since its introduction in 2001, with amendments in June 2015 reducing the target for 2020 from its previous level of 41,000 GWh per year to 33,000 GWh per year. Despite this reduction, the RET is still expected to achieve its original ambition of at least 20 per cent of Australia's electricity coming from renewable sources by 2020. State-based feed-in tariffs have also been used in the past to promote small-scale renewable generation, principally solar, by offering a generous fixed price to generators. However, the only state offering a feed-in tariff for bioenergy was Victoria and this was closed to new entrants in 2012.

In Brazil, renewable energy incentives have historically been focused on liquid biofuels rather than electricity. Brazil has been a pioneer in the use of ethanol for transport since the 1970s under its PROALCOOL programme, driven by goals of reducing oil imports and providing a market for local cane-growers (and more recently to combat climate change). Policy incentives to produce renewable electricity have not been as significant as for biofuels, largely because hydroelectricity has historically been the most economical generation option, accounting for around three-quarters of all generation. However, Brazil has more recently sought to diversify into wind and biomass generation, first by providing fixed-price subsidies (effectively feed-in tariffs) through the PROINFA scheme from 2002, before transitioning to an auction-based approach to awarding fixed-price contracts for new plants (Azuela et al., 2014).

The main source of bioelectricity in Brazil is bagasse from sugarcane, which is used for cogeneration of heat (for use in mills) and electricity. Bioelectricity accounted for 7.5 per cent of Brazil's overall electricity generation in 2014, with the country ranked fourth in the world with 32.9 TWh (terawatt-hours) of bioelectricity generated (REN21, 2015). Thus, while Brazil is more well-known for biofuels than bioelectricity, its bioelectricity sector is significant as well, being larger than Australia's in both absolute terms and as a percentage of overall electricity generation.

With regards to biofuels, Brazil requires ethanol to be blended with petrol (gasoline) at a rate of 18–27.5 per cent (Barros, 2014), varying with the availability of supply across the year. In contrast, the only biofuel mandate in Australia is in the state of New South Wales (NSW), where fuel wholesalers are required to ensure that ethanol makes up a minimum of 6 per cent of all petrol sales by volume. Plans to replace all regular unleaded petrol in NSW with E10 (10% ethanol) were cancelled in January 2012 following concerns among some motorists that E10 could cause engine damage and they would have to buy

Table 9.1 Key policy measures influencing energy cropping activities in Australia and Brazil

Category	Australia	Brazil
Positive incentives	• Renewable Energy Target (Nat) • NSW Biofuels Act (Reg – NSW only) • Excise tax rebates for ethanol and biodiesel biofuels (Nat) • Feed-in tariffs for bioelectricity (Reg – Victoria only) • Tax breaks for plantation establishment (Nat) • Grants for revegetation activities (Nat and Reg) • Sequestration credits under Emissions Reduction Fund (Nat) • Biodiversity offset trading (Reg – NSW only) • Overseas bioenergy support creating potential export markets (Int) – e.g. EU RED, US RFS • Laws protecting property rights (Nat and Reg)	• Mandated use of ethanol and biodiesel for transport (Nat) • Auction approach to funding new renewable electricity generation (Nat) • Fuel tax reductions/exemptions for ethanol and biodiesel – including greater reductions for "social fuel" (Nat) • Grants for restoration activities – e.g. World Bank, GEF (Nat, Reg and Int) • Sequestration credits under Clean Development Mechanism (Int) • Overseas bioenergy support creating potential export markets (Int) – e.g. EU RED, US RFS • Laws protecting property rights (Nat and Reg) • Access to agricultural credit for compliance with Forest Code (Nat)
Negative incentives	• Vegetation clearing laws (Reg) • Land use planning rules on plantations and forestry activities (Reg) • Industrial relations laws (Nat and Reg) • Restrictions on certain plantations earning carbon credits under Emissions Reduction Fund (Nat) • Renewable Energy Target restrictions on using native forest biomass (Nat) • Sustainability standards (Int) – e.g. EU RED,RSB, FSC, SAN	• Forest Code requiring forested land to be retained (Nat) • Agroecological zoning specific to different crops (Nat) • Industrial relations laws (Nat and Reg) • Restricting access to agricultural credit in areas of high deforestation (Nat) • Sustainability standards (Int) – e.g. EU RED,RSB,FSC, SAN • Supply-chain agreements to prevent deforestation – e.g. "Soy Moratorium" (Nat and Int)

Extension	• Natural resource management and agricultural extension services (Reg) • Plantation sector extension – e.g. Plantations2020 programme (Nat) • Energy sector extension – e.g. Regional Clean Energy programme (Reg)	• Agricultural extension services (Reg and State) • Energy sector extension – e.g. Petrobras support for smallholders growing biodiesel feedstock (Nat)
Technology change	• Renewable Energy Target encourages innovation (Nat) • RandD schemes – e.g. Biofuels Capital Grants Scheme and Second Generation Biofuels Research and Development Program (Nat) • Financing programmes – e.g. Clean Energy Finance Corporation (Nat)	• RandD/financing schemes – e.g. National Bank for Social and Economic Development • Global financing – e.g. World Bank, REDD++
Knowledge and values gathering	• Environmental impact assessment processes (Reg) • Research into impacts of energy crops on land – e.g. Oil Mallee (Nat and Reg)	• Research into impacts of multifunctional production systems (Nat)
Broader institutional arrangements	• Division of responsibilities between Commonwealth and states under constitution (Nat) • International conventions – e.g. Kyoto Protocol Commitments (Int) • Endorsement of RSB standards as sustainability standard under NSW Biofuels Act (Reg)	• Division of responsibilities between Union, states and municipalities under constitution (Nat) • International conventions – e.g. Ability to participate in the Clean Development Mechanism under the Kyoto Protocol (Int)

Note: Int = international policy measure, Nat = national policy, Reg = policies specific to individual states or regions within each country.

more expensive premium fuels. Biodiesel use is much lower than ethanol in both countries, but again Brazil's usage is much higher than Australia's. The NSW mandate for biodiesel is 2 per cent of all diesel sales, with plans to increase to 5 per cent postponed indefinitely in December 2011 due to a lack of local supply. In contrast, Brazil's national mandate for biodiesel was 7 per cent as of November 2014 (REN21, 2015).

Ethanol and biodiesel are also given an advantage over their fossil fuel competitors in both Australia and Brazil through the fuel taxation systems. Up until 2015, the Australian Government paid grants and rebates to biofuel producers to fully offset the fuel excise charged on ethanol and biodiesel. From July 2015, these programmes were ended and excise will gradually be phased in, starting at zero and rising by 2.5 cents per litre each year to a maximum of 12.5 cents per litre for ethanol by 2020 (Australian National Audit Office, 2015). Over time, this will reduce the tax breaks afforded to biofuels, but they will still attract a lower rate than petrol and diesel (roughly 39 cents per litre in 2015).

A similar system of reduced fuel tax is employed in Brazil, with ethanol effectively tax-free at present and biodiesel attracting differing levels of tax reduction depending on the source of the feedstock (Barros, 2014). As discussed in Chapter 6, the use of the social fuel label for certain types of biodiesel is designed to encourage the sourcing of feedstocks from poor family farmers in the poorer north and north-east regions of Brazil. Social fuels are also given preference under the government-run biodiesel auctions that occur every two months, which help provide price certainty for suppliers (Barros, 2014).

Extension and technology change have also played an important role in the development of energy cropping in Australia and Brazil. Extension services may be provided by natural resource management agencies, the agriculture sector or by energy suppliers. The role played by Brazil's state-owned oil company Petrobrás in supporting small family farmers is a good example of extension provided by buyers of energy crops towards feedstock producers (see Chapter 6). In Australia, energy companies have partnered with state government agencies on mallee trials in Western Australia (Verve Energy, 2006) and NSW (Delta Electricity, 2010). NSW also has a system of Renewable Energy Precincts where government-funded coordinators provide extension services to community members interested in renewable generation, but to date this has mostly been based around wind energy.

Research and development support has played a major role in Brazil becoming a global leader in biofuel production and this continues with the focus on advanced biofuels under programmes such as "Paiss Agricola" funded by the National Bank for Social and Economic Development (Barros, 2014). The Australian federal government has also funded research into advanced biofuels under the Second Generation Biofuels Research and Development Program. The development of flex-fuel vehicles that can run on any combination of ethanol and gasoline was a key development in increasing ethanol use in Brazil, with these vehicles continuing to be supported through preferential tax treatment (Barros, 2014).

Apart from policies specific to renewable energy, Table 9.1 also highlights the importance of policies related to biodiversity conservation, restoration and social issues in both Australia and Brazil. As highlighted in Chapter 8, Australia has been a global innovator in areas such as carbon sequestration offsets, biodiversity offsets and biodiversity auctions. The BushTender auction process in the state of Victoria seeks to promote cost-effectiveness in the spending of state government money. Following a different market-based approach, the BioBanking scheme in NSW allows developers to use offsets for environmental damage they wish to cause, which increases flexibility as well as providing an alternative to government funding for biodiversity protection and restoration activities.

As discussed in Chapter 8, Australia no longer has a national emissions trading scheme, but has retained the framework of the Carbon Farming Initiative by rolling it into the new Emissions Reduction Fund (ERF). This allows revegetation and farm forestry projects to sell their verified sequestration credits to the federal Government via an auction scheme. This may allow plantations of energy crops to earn credits for the long-term average sequestration that is held within the plantations. As discussed in Chapter 4, the level of carbon in a plantation will be at a high just prior to harvest and at a low just after harvest, with the average falling somewhere in between.

While the approved methodologies relating to farm forestry under the Carbon Farming Initiative could conceivably allow energy cropping projects to supplement their earnings with carbon credits, there are potential barriers to this occurring. For example, plantations are ineligible to earn carbon credits if they were funded by a managed investment scheme (which provides tax breaks for establishment costs) or if they are in high-rainfall areas (>600 mm yr^{-1}). This latter condition is due to concerns around competition for water and represents an attempt to protect food production and other forms of agriculture from competition with carbon plantings.

Brazil does not have a national emissions trading scheme (or emissions reduction auction like the ERF), but it does have access to the Clean Development Mechanism of the Kyoto Protocol, which allows reforestation projects to earn credits for sale to Annex I countries. Eligible sites must have been non-forested as of 1989. Brazil has three reforestation/afforestation projects registered under the CDM, including the Plantar Group project discussed in Chapters 4 and 6 that is specifically designed around the production of woody crops for energy (UNFCCC, 2010). This project involves the use of wood from eucalyptus plantations in the state of Minas Gerais to produce charcoal for iron production. As discussed in Chapter 6, this project has been criticised by NGOs such as Carbon Market Watch (2010) for having negative social impacts. However, the plantations in question have been certified by the FSC, which includes criteria on land rights and community relationships. This highlights that, even where a project has multiple layers of oversight (i.e. Brazilian authorities, CDM and FSC), some stakeholders may still feel that the negatives of the project outweigh the positives.

When it comes to providing protections for key environmental and social values in Australia and Brazil, negative incentives have a critical role to play. Most of the negative incentives shown in Table 9.1 are not specific to bioenergy but relate to a range of possible land use activities. Australia saw a number of new state-based laws introduced in the mid-2000s to reduce the rate of land-clearing, particularly in Queensland and New South Wales. In the case of Brazil, deforestation has been a major global controversy, particularly in the Amazon basin. As discussed in Chapter 3, Brazil has been successful in dramatically reducing its rate of deforestation in recent years through a combination of government regulations (including reforms to its Forest Code) and interventions by key stakeholders in the soy and beef supply chains. Brazil's use of agro-ecological zoning was also discussed in Chapter 3, including different treatment for sugarcane, which cannot move into previously deforested areas, and oil palm, which is actively encouraged in such areas to reinstate some form of perennial tree cover.

International pressure and supply-chain interventions often play a greater role in providing environmental and social protections in developing countries, especially where national governments are unwilling or unable to provide adequate protection through regulation. In some cases, this may take the form of an agreement between key supply chain participants who can exert significant influence over suppliers, as demonstrated by the role that Brazil's "Soy Moratorium" has played in slowing Amazon deforestation (Gibbs et al., 2015). In other cases, global certification schemes such as those operated by the FSC, RSB or Rainforest Alliance may help to promote changes at the farm level. However, such certification schemes are by their very nature best suited to maintaining rather than enhancing conditions (e.g. protecting forests rather than incentivising restoration). Also, as highlighted by the example of the FSC-certified eucalypt plantations in Minas Gerais, there is no guarantee that such schemes will be seen as adequate by all stakeholders.

Analysing policy: goals, issues, trade-offs and synergies

Figure 9.2 attempts to map the different goals underpinning the various policy measures relating to renewable energy, environmental protection, restoration and socio-economic development in Australia and Brazil. On the left-hand side of the figure is a list of broad sustainability issues that feature in the policy measures discussed in the previous section. The figure shows how these broad sustainability issues may be translated into more specific and variable goals across global, national, regional and local scales. The lines marked S, T and S/T highlight some of the potential synergies (S) and trade-offs (T) that can exist between the various goals, which are explained further in Table 9.2.

Environmental issues such as biodiversity conservation, soil health and climate change are topical at all scales, but the emphasis can vary. Global goals around biodiversity and soil health, such as those the contained in international conventions or sustainability certification standards, generally focus on maintaining existing conditions (i.e. not making things worse). However, there may also be local opportunities for land uses such as energy cropping to contribute to active enhancement, such as habitat provision from eucalyptus or oil palm crops in Brazil or salinity mitigation from mallee crops in Australia. Similarly, climate change goals can also vary, with mitigation being most dominant at the global and national scales, and adaptation requiring more context-specific actions at the national, regional and local scales. Energy crops may also play differing roles in mitigating climate change at different scales, helping to meet targets for renewable energy at the national scale, while contributing to carbon sequestration at the local scale.

Renewable energy development is underpinned by goals such as energy security, climate change mitigation and socio-economic development. However, these goals may vary over time and space, with Brazil's development of a world-leading ethanol industry being initially driven by a desire to reduce oil imports (i.e. national-scale energy security) and pre-dating concerns about climate change. Compared with these national-scale concerns, the local-scale goals of a landholder contemplating energy cropping are more likely to be based on goals such as obtaining a livelihood, being able to stay on the land or increasing the resilience of their farming system. Brazil's biodiesel support programme provides an example of how a national-scale programme aimed at increasing renewable energy production can also be tailored to promote local-scale objectives relating to livelihoods and social equity.

Social issues such as land rights, job creation and food security are topical from the global through to local scales, but may differ in nature across these scales. National (and state/provincial) governments play an important role in setting workplace conditions and protecting land rights. Having secure land rights is clearly an important goal for local landholders, with the reforms to Brazil Forest Code discussed in Chapter 5 showing how security of land rights can be used as incentive for compliance with forest protection laws.

Maintaining food security on a global scale has been a major issue around the expansion of biofuels from common agricultural crops, but this is a very challenging goal to achieve through local or even national-scale policies. Australia and Brazil are both major agricultural exporters, but have to balance the goal of feeding the world against maximising export opportunities, which may come from fuel rather than food. Similarly, at a local scale, land use flexibility may be a more important goal for landholders than ensuring that the crops they produce contribute most to global food security – a point that was highlighted in the landholder interviews cited in Chapter 6 from the central west of NSW.

Where synergies exist between different goals, policy-makers may be able to promote multiple outcomes with a single policy measure (e.g. reducing GHG emissions to simultaneously prevent climate change and protect biodiversity). Where trade-offs exist, policy-makers need to be aware that the promotion of one outcome may have negative consequences for other goals. For example, land

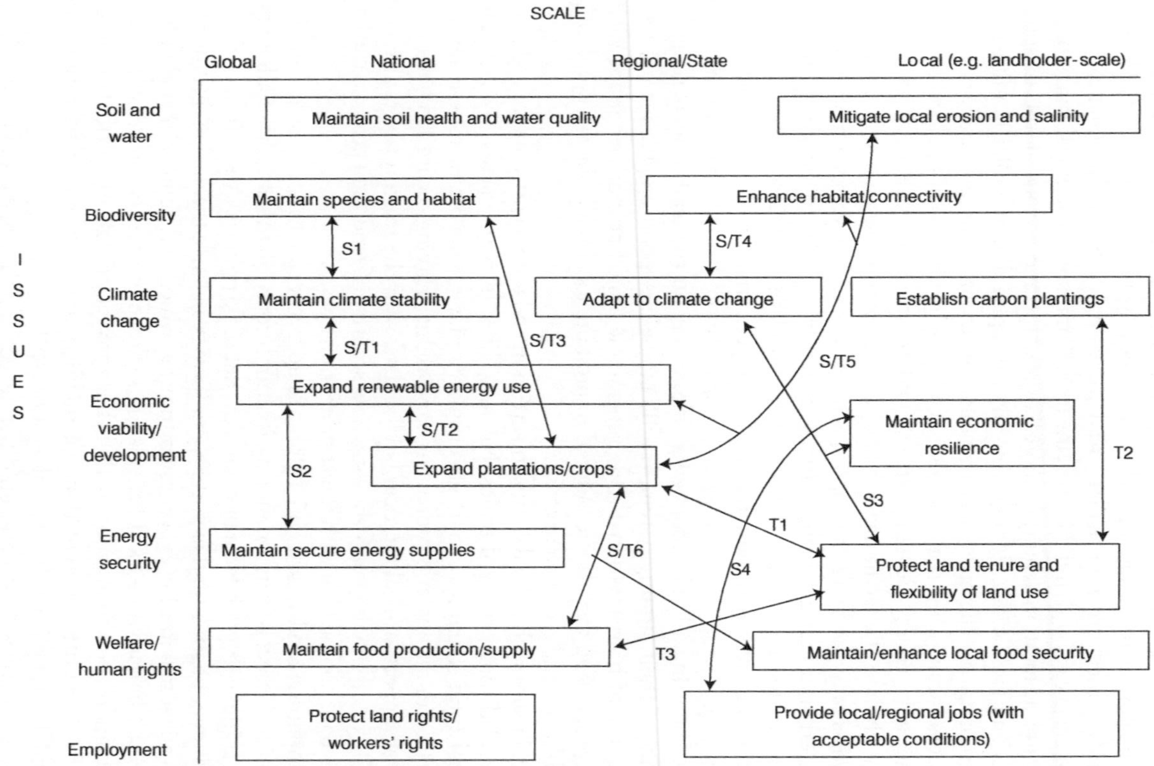

Figure 9.2 Goals underlying existing policy measures, matched to their corresponding sustainability issues (vertical axis) and scale (horizontal axis). Lines with arrows show potential synergies (S), trade-offs (T) and mixtures of synergies and trade-offs (S/T)

Table 9.2 Synergies and trade-offs between various goals from Figure 9.2

Goal 1	Goal 2	Synergy or trade-off	Explanation
Maintain climate stability	Maintain species and habitats	Synergy (S1)	Climate change poses a major threat to biodiversity
Expand renewable energy	Maintain energy security	Synergy (S2)	Renewable energy is capable of both
Maintain land use flexibility	Adapt to climate change and maintain economic resilience	Synergy (S3)	Flexibility of land use is critical to adaptive capacity and resilience
Provide local jobs	Maintain economic resilience	Synergy (S4)	Local job provision is a key component of economic resilience
Expand energy crops	Maintain land use flexibility	Trade-off (T1)	Woody energy crops often have longer wait times to initial harvest than annual crops
Establish carbon plantings	Maintain land use flexibility	Trade-off (T2)	Carbon plantings require long-term sequestration (often defined as 100 year commitment)
Maintain food production	Maintain land use rights/ flexibility	Trade-off (T3)	Restricting the use of land to maintain food production reduces the options available to landholders (e.g. energy cropping).
Expand renewable energy	Maintain climate stability	Either/both (S/T1)	Bioenergy may have high or low life-cycle emissions depending on how it is produced (e.g. if deforestation is involved)
Expand plantations	Expand renewable energy	Either/both (S/T2)	Plantations may be used for energy or other products (e.g. timber, paper)
Expand plantations	Maintain climate stability	Either/both (S/T3)	Plantations can sequester carbon, but not if forests are not cleared for establishment
Enhance habitat and connectivity	Adapt to climate change	Either/both (S/T4)	Connectivity facilitate species migration, but if landholders lose their income from the land, social adaptation may be hampered
Expand energy plantations	Enhance habitat connectivity, erosion control and salinity mitigation	Either/both (S/T5)	Plantations can contribute to these goals, but can also have negative impacts if planned poorly
Expand energy plantations	Maintain food production	Either/both (S/T6)	Food crops may be diverted to biofuels, but carefully integrated plantations can enhance food production (e.g. Oil Mallee in WA)

use restrictions to ensure food security may inhibit the flexibility of land use and hence the ability of local landholders to respond to changing climatic or market conditions.

The last set of relationships described in Table 9.2 involve situations where either synergies or trade-offs may be encountered, depending on how policy actions are carried out. These cases are perhaps the most common and require careful policy design and assessment of potential impacts. For example, measures to promote renewable fuels may assist with climate change, but only if eligibility is restricted to fuels with low life-cycle GHG emissions. Similarly, expansion of energy crops may come at the expense of food production, but there are also cases where it can enhance food production by protecting soil health or providing food as a co-product.

Policy problems and gaps in Australia and Brazil

After analysing the relationships between existing policy conditions, issues and goals, the next step in developing effective sustainability policy is to identify a set of *policy problems* to be solved. In the words of Stephen Dovers, the difference between issues and problems is that "issues are for being concerned about and debating, problems are for resolving" (Dovers, 2005, pp. 41–2). Table 9.3 highlights a number of key policy problems that remain around Australian and Brazilian energy cropping. These are divided into three categories relating to barriers to sustainable energy cropping, threats from unsustainable energy cropping and uncertainty. This is not designed to be an exhaustive list, but rather a set of priority areas where policy reform could be undertaken.

Many of the policy problems highlighted in Table 9.3 are common to both Australia and Brazil. However, there are some notable differences. Australia faces greater challenges in terms of energy crop competitiveness, while Brazil already has a highly competitive sugarcane ethanol industry and its remaining challenges relate more to biodiesel and advanced biofuels. Brazil also lacks a national scheme to earn carbon credits from sequestration and has a greater reliance on international sustainability standards and supply-chain agreements. In terms of risks, Brazil faces greater challenges around continued deforestation, security of land rights and labour rights, while Australia faces greater uncertainty around landholder attitudes and commercial viability due to a lack of experience with energy cropping.

Potential policy changes

This section outlines a number of potential policy solutions that could be applied in Australia and Brazil to address the policy problems outlined in Table 9.3. These options draw on the lessons learnt within each country, as well as lessons learnt from experience with policy measures undertaken in places such as the EU and the US. The potential options are explored under the three themes shown in Table 9.3 covering barriers, threats and uncertainty.

Table 9.3 Policy problems to be resolved for the case studies

Theme	Policy problem	Relevance to: Australia			Brazil		
		L	M	H	L	M	H
Barriers to the production of sustainable energy crops	Many forms of bioenergy are more expensive than fossil fuels			X		X	
	Energy cropping not competitive with other land uses			X	X		
	Energy cropping with restoration co-benefits is more costly to implement			X			X
	Lack of incentives to locate or design plantations for benefits to biodiversity or soils			X			X
	International sustainability standards and supply-chain agreements incentivise maintenance but not restoration	X				X	
	Barriers to earning carbon credits from sequestration by energy cropping systems	X					X
Threats from the expansion of unsustainable energy crops	Energy crops may contribute to deforestation and other environmental damage	X				X	
	Energy crops may pose social threats such as land grabs and exploitation	X				X	
	Energy crop expansion may pose a threat to local, national or global food security		X			X	
Uncertainty	Uncertainty around whether advanced biofuels will become competitive with fossil fuels			X			X
	Uncertainty around landholder attitudes and risks related to growing energy crops			X	X		
	Uncertainty about habitat value and soil impacts of energy crops in different contexts			X			X
	Uncertainty around the environmental and social risks from unsustainable energy cropping		X				X

Overcoming barriers for sustainable energy crops

The different circumstances facing energy crops in Australia and Brazil require targeted policy measures aimed at overcoming barriers in each country. In Australia, the lack of established energy crops, the greater significance of bioelectricity relative to liquid biofuels and the lack of connections with other bioenergy markets necessitate a focus on domestic competitiveness, at least in the short to medium term. This relates to the competitiveness of energy crops with established fossil fuel technologies, as well as with other forms of bioenergy based on wastes (e.g. bagasse and wood waste for electricity, waste wheat starch and beef tallow for biofuels). In Brazil, there are also opportunities to expand domestic use of energy crops, but the use of biofuels for transport is already high and the use of fossil fuels for electricity is low. These factors, combined with stronger connections to overseas biofuel markets, make the question of competitiveness in Brazil a much more international one.

In Australia, the Renewable Energy Target acts as the main option for enhancing the competitiveness of bioenergy relative to coal and other fossil fuels within the electricity sector. However, it features two main flaws in relation to energy crops. First, it offers little certainty for those investing in energy crops today, as its targets are only set until 2020 and the numerous reviews and changes in its 15 year history have not led to a stable investment environment for renewables. Second, it offers no advantages for energy crops over more established technologies like wind or landfill gas.

The first problem could be addressed by re-introducing a carbon price that would enhance the competitiveness of renewable technologies by making fossil fuel generation more expensive. As discussed in Chapter 8, Australia had a fixed carbon price (i.e. a carbon tax) from 2012 to 2014 and this was due to transition to a cap-and-trade scheme (similar to the EU Emissions Trading Scheme) before it was scrapped by the incoming conservative government. Under the previous plan, the RET would act as the main driver of renewable energy investment until 2020, after which the carbon price would provide investment certainty for renewables, especially as it was expected to increase over time as emissions caps were progressively tightened (Treasury, 2011).

Reintroducing an emissions trading scheme could help to improve the competitiveness of renewables against fossil fuels over the medium to long term and would comply with the "polluter pays" principle under which the costs of environmentally damaging activities are internalised to those undertaking them. It could also enhance the ability of energy croppers to earn carbon credits from sequestration by providing access to a variety of buyers rather than the current arrangements under the Emissions Reduction Fund where the federal government is the only buyer of sequestration credits.

While a carbon price would improve the competitiveness of renewable energy in Australia generally, it has certain limitations. A carbon price is not capable of supporting emerging technologies such as energy crops through their early development phase, nor can it properly reward energy croppers for the ecosystem benefits they might provide for soil health or biodiversity. A modified version of the RET or the use of feed-in tariffs would provide a better option for achieving these goals. The RET could be modified to incorporate the type of "banding" system discussed in Chapter 8 (with the UK given as an example), whereby energy crops are awarded more certificates per MWh than more established forms of renewable energy. In addition to promoting emerging technologies, such an approach could also be used to preference forms of renewable generation that provide ecosystem benefits for soils, waterways or biodiversity (i.e. energy crops that provide these benefits get more certificates than energy crops that do not).

An important consideration mentioned in Chapter 8 regarding banding and Australia's RET is that a form of banding was experimented with in the past, when a "multiplier" was awarded to small-scale solar generation. While this system encouraged increased installation of rooftop solar, it was criticised for flooding the market with "phantom" certificates that came at the expense of other forms of renewable generation (Buckman and Diesendorf, 2010). This led to the current arrangement, where no banding is applied under the RET, but small-scale generation is supported under a separate sub-mandate (the Small-Scale Renewable Energy Scheme). To avoid a repeat of these conflicts, a possible future pathway after 2020 would be to use a carbon price to promote mainstream renewable technologies such as wind, while a modified RET is used to preferentially support emerging renewable technologies. This modified RET could apply banding for energy cropping systems that offered ecosystem benefits, such as mallee energy crops that help mitigate salinity, but without affecting the incentives around established renewables such as wind.

Feed-in tariffs could be used in a similar way to provide targeted fixed-price support for energy crop generation, such as the bonuses that have been given for certain feedstocks in Germany, also discussed in Chapter 8. The chief advantage of using feed-in tariffs over a modified RET with banding would be the investment certainty offered by the fixed prices of a feed-in tariff. The disadvantages are that feed-in tariffs in Australia are the responsibility of state governments rather than the federal government and there is a history of several states offering generous feed-in tariffs for solar which then had to be modified as solar installation costs fell. In NSW this even included a threat (later withdrawn) to retrospectively reduce tariffs for participants who had signed up at the original higher rate under 20-year contracts (Solar Choice, 2011). The risks of inconsistency between states and unexpected changes to tariffs reduce the potential for investment certainty under feed-in tariffs.

In Brazil, electricity generation from energy crops could potentially benefit from the use of a mandate scheme that featured banding or feed-in tariffs. These approaches could help to overcome one of the challenges with the auction system currently in place – that it awards contracts for proposed generation which may be delayed or not delivered at all. However, the advantage of the auction system currently in place is that it reduces risk for risky investments such as energy crop generation, by providing investors with certainty of both price and quantity (Azuela et al., 2014). If the auction system is retained, one option for preferring energy crops would be to have technology-specific auctions. These have been used previously to promote certain types of generation (e.g. renewables only) and this principle could be extended to ensuring a certain number of energy-crop-only auctions were held each year.

Carbon pricing is likely to play a less important role in promoting energy crops in Brazil than in Australia, at least in the near future. As Brazil does not have binding emissions targets under the Kyoto Protocol, the Clean Development Mechanism is likely to remain the leading opportunity to supplement income from energy cropping with carbon credits. These credits may come from either emissions reductions below a baseline (e.g. using energy crops to replace fossil fuels in local processing or electricity generation) or from carbon sequestration (e.g. average level of carbon stored in energy crops across harvest cycles). Both options are possible, as shown by the Plantar eucalypt-charcoal example from Minas Gerais. However, there are challenges around getting projects through the CDM process, as well as competition from low-cost abatement options in other developing countries. This highlights the need to complement carbon trading with other policy measures aimed at preferring energy crops with associated co-benefits (such as technology-specific renewable energy auctions).

In relation to liquid biofuels, Brazil has historically been an important policy innovator and has the potential to contribute to continued innovation around advanced biofuels such as cellulosic ethanol and renewable diesel (i.e. from thermochemical conversion of biomass). Barros (2014) reports that, by 2016, Brazil is expected to have two demonstration plants for advanced biofuels and three commercial plants in operation. However, the extent to which Brazil develops these fuels is also likely to depend on developments in the US and the EU.

The US and EU have ensured that advanced biofuels are preferentially promoted under their biofuel mandates by requiring progressive increases in life-cycle GHG savings requirements (US) and "double-counting" for cellulosic fuels (EU). In contrast, Brazil's ethanol and biodiesel mandates do not preference advanced biofuels over their first-generation counterparts. Similarly in Australia, neither the fuel tax arrangements at the federal level nor the mandates in the state of NSW feature any differentiation between advanced and first-generation fuels.

Globally, the development of advanced biofuels has been slower and more expensive than expected, with over-ambitious US targets having to be revised downwards (Environmental Protection Agency, 2013). However, as breakthroughs are achieved in the US and EU, these could benefit Brazil and Australia in two main ways. Firstly, the US and EU markets could become a destination for feedstock supply, particularly in the case of Brazil which has the potential to produce biomass from eucalyptus energy crops more cheaply than North American feedstock producers (Stephen et al., 2013). Secondly, technological breakthroughs from the US and EU may be applied in Brazil and Australia to increase biofuel production domestically. Again, Brazil is more likely to benefit from this than Australia, for example by combining overseas breakthroughs around enzymes for cellulose conversion with domestic research into locally available feedstocks such as sugarcane bagasse and pre-existing distribution networks for ethanol.

Technological breakthroughs may be essential to the development of advanced biofuels, but these will not be sufficient to promote energy crops over other sources of biomass, such as bagasse and other agricultural residues. However, energy crops could be preferred through differentiated support under domestic biofuel policies in Australia and Brazil. Brazil already offers some form of differentiation between feedstocks under its fuel tax arrangements for biodiesel, whereby feedstocks sourced from small family farmers in poorer regions (i.e. "social fuel") attract higher tax exemptions. These tax advantages could be extended to cellulosic energy crops, particularly those that offer benefits for soils or biodiversity (i.e. an "ecosystem fuel").

While Brazil's taxation arrangements for biodiesel could be modified to promote advanced fuels and "ecosystem fuels", this is more difficult for ethanol due the fact that all ethanol in Brazil already has zero fuel tax at present (Barros, 2014). The Australian situation is complicated by the recent decision to remove rebates on ethanol and biodiesel, but it would still possible to phase in a higher level of fuel tax (excise) for first-generation biofuels than for advanced biofuels. These issues are highly political in both countries, with different stakeholders lobbying for preferential treatment. However, if both countries were able to undertake a comprehensive review of fuel taxation, it is conceivable that systems could be put in place whereby the level of tax exemption was dependent on whether a biofuel represented an emerging technology, whether it qualifies as a "social fuel" (e.g. with benefits for small family farmers) and whether it qualifies as an "ecosystem fuel" (e.g. with benefits for soils or biodiversity).

As with fuel taxation arrangements, it is also possible to structure biofuel mandate schemes to preference emerging fuels and those with social or environmental benefits. One option is to set a series of sub-mandates, with fuel suppliers having to ensure that, for example, 5 per cent of their diesel is from renewable sources and at least 10 per cent of that qualifies as an "ecosystem fuel". Alternatively, an approach similar to that of the European Union's Renewable Energy Directive could be employed, whereby emerging fuels are promoted through "double-counting" or even "quadruple-counting". This would require a certificate-based system such as the UK Renewable Transport Fuel Obligation (RTFO), with fuel suppliers having to surrender a set number of certificates to the government, which they obtain by providing certain renewable fuels (e.g. 1 certificate per litre for general biofuels, 2 per litre for a social fuel, 4 per litre for a fuel that is both a social and an ecosystem fuel).

Moving beyond bioenergy-specific policies, it is also important that incentives and extension services are provided at the landholder level. These incentives could include grants, tax breaks or low-interest loans to help overcome establishment costs for energy crops, with this option being favoured over price support by landholders interviewed in the central west of NSW (Baumber et al., 2011). This may require a significant degree of institutional change, particularly in Australia, where the government agencies responsible for providing restoration grants are often different to those providing loans and tax breaks for biofuels, forestry or agriculture. For example, significant reforms would be required to programmes such as BushTender in Victoria or BioBanking in NSW to enable commercial energy cropping systems to be eligible for the payments for ecosystem services that these schemes provide. The Carbon Farming Initiative (now part of the Emissions Reduction Fund) could provide a model for integrating conservation and production goals, as it includes methodologies for both environmental plantings and farm forestry activities.

Apart from funding for energy crop establishment, extension services are likely to play a critical role in helping landholders engage with energy cropping as a new land use option. The role played by Petrobrás in helping small family farmers grow feedstocks under Brazil's biodiesel support programme (cited in Chapter 6) highlights how on-ground extension services can complement higher-level policies aimed at creating a market or supporting prices. Among the extension services that have been provided by Petrobrás for biodiesel feedstock growers are supply of high quality seeds, technical support, purchasing feedstocks at above-market prices, helping to integrate biofuel feedstocks with food production and working with local NGOs to ensure the fairness of contracts (Lima, 2012).

The Australian energy sector does not have the same degree of state ownership as in Brazil and it is unlikely that an energy company would perform the same role as Petrobrás in relation to supporting energy croppers. However, electricity companies have partnered with government agencies and landholder organisations to engage in research and development around energy cropping in the past, particularly around mallee eucalypts in Western Australia (Verve Energy, 2006) and NSW (Delta Electricity, 2010). These types of partnerships could be further advanced through institutional and regulatory relationships that bring together the roles of different stakeholders. Under these partnerships, government agencies could help support innovation and provide payments for ecosystem services, energy companies could help establish secure relationships for reliable fuel supply and landholder organisations could help to coordinate landholders, such as the Oil Mallee Association in WA, which has a grower base of over 1200 (URS Australia, 2009), or the Lachlan Renewable Energy Alliance around Condobolin in NSW. Energy companies may participate voluntarily or could be compelled to engage with landholders under new obligations design to make up for some of the environmental damage caused by their fossil fuel use (e.g. protecting soils and biodiversity as a form of climate change adaptation).

Mitigating threats from unsustainable energy crop development

As discussed previously, there are already a range of policy measures in place in Australia and Brazil to protect against the threats posed by unsustainable land use activities, such as laws preventing land-clearing and protecting land and labour rights. The policy suggestions discussed here are in addition to these existing protections.

When it comes to regulatory restrictions, a key issue is how they should be targeted. While some measures may cover a broad range of land uses or production systems, others may be targeted at a single land use or product, such as bioenergy. A general principle followed here is that regulatory restrictions should be applied broadly across all production systems (e.g. food, fuel, fibre), unless a particular product poses a unique threat or has specific characteristics that require targeted regulations. While energy cropping could pose threats such as land clearing, land degradation and land use competition, it is not unique in posing these threats. As such, it is recommended that broad-based land use regulations are maintained, such as the various state native vegetation clearing laws in Australia and the Forest Code in Brazil.

In relation to competition for land, there is no convincing argument for imposing restrictions on bioenergy-based land uses alone, as other activities such as plantation forestry, cotton-growing, mining and urban expansion can also have indirect impacts on food production and prices. However, it may be prudent in some cases to reserve areas of land for particular purposes, such as food production. This would not act as a restriction on energy cropping alone, but would require any non-food land use to undergo an assessment to determine whether it was likely to have an impact on food security at a local or national scale. The Australian state of NSW has experimented with planning policies of this nature, employing a Strategic Regional Land Use process to identify areas that could be at risk from mining or coal seam gas development (Department of Planning and Infrastructure, 2012). While the focus is currently on mining and coal seam gas and some groups have rejected the proposed protections as inadequate (e.g. NSW Farmers' Association, 2012), this broad framework could form the basis of an integrated planning approach that ensures adequate food production while still allowing for the development of new industries such as energy cropping. Implementation would ideally be on a trial basis, with extensive community consultation to determine appropriate land use values and principles.

An alternative (or complement) to restricting non-food land uses is to vary biofuel and bioelectricity mandates at times of high food prices or crop shortages. Brazil already employs a variable ethanol blending mandate that takes into account seasonal fluctuations in ethanol supply. The NSW biofuel mandates in Australia have been kept on hold for a number of years now due to inadequate supply. This shows the potential to use these mandates in an adaptive fashion to respond to prevailing conditions around supply, demand and pricing.

In the case of Brazil, international certification schemes and supply chain agreements are likely to continuing playing a much greater role in relation to energy cropping than they do in Australia. This is due to the ongoing nature of Amazon deforestation, the high level of international interest in this issue and the importance of exports to the biofuel sector. Certification of individual feedstock producers by the RSB or other bodies may be sufficient to satisfy certain markets, such as the biofuel sustainability criteria imposed by the EU. However, in other cases it may be necessary for large players in the supply chain to band together (as under the "soy moratorium") to collectively ensure that the energy cropping sector is not linked to deforestation. For example, should Brazil become a major exporter of biodiesel or cellulosic ethanol from woody crops produced in or near the Amazon basin, it may be necessary to create a "biofuel moratorium" to satisfy overseas markets that these biofuels are not contributing to deforestation.

International sustainability standards are less relevant for Australia, but may be important if the level of biofuel or feedstock export increases in the future. As such, care should be taken to ensure that Australia's regulatory regime is consistent with the latest developments around international sustainability schemes. One measure that could be adopted in both Australia and Brazil is the use of life-cycle greenhouse gas benchmarks for biofuels, following the approach taken by the EU, US and RSB. The governments of Australia and Brazil could support biofuel producers in accessing overseas markets by providing standardised methodologies and default values that enable fast, low-cost assessment of life-cycle GHG emissions.

Dealing with uncertainty

There are four main sources of uncertainty affecting the development of sustainable energy crops in Australia and Brazil: uncertainty around the commercial viability of energy cropping, uncertainty around landholder attitudes to growing energy crops, uncertainty around the environmental benefits of sustainable energy cropping and uncertainty around the environmental and social risks from unsustainable energy cropping. The range of policy options that could be employed to deal with these sources of uncertainty cut across the policy categories of positive incentives, negative incentives, technology change and knowledge and values gathering. Furthermore, when implementing the policy measures suggested in the previous sections on barriers and threats, an adaptive management approach is required, involving trials, monitoring of outcomes, targeted research to address knowledge gaps and the flexibility to adjust policy parameters in response to new information.

The commercial viability of many types of energy cropping is uncertain, particularly cropping systems that trade off some degree of production efficiency in return for ecosystem enhancements for soils, waterways or biodiversity. The incentives and extension measures cited in the section on overcoming barriers help to address this uncertainty by enabling "learning by doing" within supported markets, something that was highlighted in Chapter 7 as a key element in the development of other energy crops, such as corn ethanol in the US. However, there is also a need for basic research and the development of new technologies. As discussed previously, this research and development is not exclusive to either Australia or Brazil but includes the application of technological breakthroughs made in places such as the US and the EU.

Overcoming uncertainty around landholder attitudes to energy cropping is primarily a domestic concern within both Australia and Brazil. Social research has the capacity to reduce this uncertainty by illuminating the objectives and barriers for specific landholder groups. Examples of the information that can be provided by social research of this nature has been shown through the various examples cited in this book relating to mallee eucalypts in Australia (Baumber et al., 2011), biodiesel feedstocks in Brazil (Lima, 2012), short-rotation coppicing in the UK (Dockerty et al., 2012) and jatropha in Africa (Romijn et al., 2014). Trials are also necessary to identify where the barriers and opportunities lie.

The potential benefits of energy cropping for ecosystem health are also highly uncertain. If there is to be effective differentiation under renewable energy support schemes between energy crops that enhance ecosystem services and those that do not, systems will need to be developed to appropriately measure and verify these impacts. Existing certification schemes, such as those of the RSB or FSC, do not adequately perform this task as they are based primarily on the maintenance of existing ecosystem services rather than their enhancement. However, lessons may be learnt from other international mechanisms that have been developed for measuring and verifying ecosystem services.

In relation to carbon sequestration, the methodologies used by the Clean Development Mechanism and Australia's Emissions Reduction Fund may help to assign sequestration values to energy cropping systems. The carbon savings methodology used by the RSB also allows for sequestration benefits to be factored into life-cycle GHG calculations – and the state of NSW has endorsed the RSB standards as a recognised sustainability standard under its biofuel mandate programme. One pathway forward would be to be use minimum GHG savings based on the RSB or EU methodologies to determine basic eligibility for a bioenergy support programme (e.g. mandate or tax break). In addition to these minimum benchmarks, a higher level of support (e.g. double-counting under a mandate or greater tax breaks) could be provided to biofuels that have a GHG saving greater than 100 per cent due to carbon sequestration (i.e. they are carbon-negative). This approach would benefit from the work done by the RSB and EU in reducing uncertainty around biofuel life-cycle emissions.

The benefits of energy crops for biodiversity are harder to measure and model than carbon savings, but there are also relevant examples to draw on, such as BushTender in Victoria and BioBanking in NSW. Costa Rica has also developed metrics for watershed protection and landscape beauty within its framework for payments for ecosystem services (Porras et al., 2013). In relation to soils, efforts to quantify the reversal of land degradation have attracted attention lately with the setting of international goals around "zero net land degradation" (Tal, 2015). This may also result in metrics that can be applied to the soil health benefits of energy cropping systems. Existing tools such as Landscape Function Analysis have also been used to demonstrate the benefits of some energy crops in Australia for soil stability, water infiltration and nutrient cycling (Baumber, 2012).

The final source of uncertainty relates to the risk of environmental damage from unsustainable forms of energy cropping. The regulatory and certification measures discussed previously are one way of dealing with this uncertainty. This approach is consistent with the precautionary principle discussed in Chapter 1, which states that "where there are threats of serious or irreversible damage, lack of full scientific certainty shall not be used as a reason for postponing cost-effective measures to prevent environmental degradation" (United Nations Conference on Environment and Development, 1992, Principle 15).

The other important policy response to risks of negative social and environmental impacts is to reduce the level of uncertainty by supporting research into the potential impacts of a range of energy cropping systems. As shown by examples such as the Plantar eucalypt plantations in Minas Gerais, there will inevitably be differing perspectives on what level of risk is too high and what level of scientific certainty is required before proceeding with a new land use option. As such, research into these issues should also be aimed at understanding the values and tolerance of risk held by various stakeholders towards factors such as habitat protection, food production and maintenance of existing social structures.

Conclusion

This chapter has explored the existing policy environments within Australia and Brazil and suggested some potential areas of policy reform to promote the most sustainable forms of energy cropping while mitigating the risks of unsustainable energy crops. Both countries have sought to create a policy environment that encourages the use of renewable energy and protects against environmental destruction and the exploitation of vulnerable people. Brazil has been more innovative than Australia in relation to energy cropping, as shown through its pioneering development of ethanol mandates starting in the 1970s, as well as its more recent innovations designed to support small family farmers under its biodiesel support programme. While Australia has not been a leader globally in relation to energy cropping, it has shown significant policy innovation around payments for ecosystem services, including tradeable credits for carbon sequestration and biodiversity and auction-based approaches that measure and reward habitat protection.

Moving forward, each country could benefit from looking at what the other has done in relation to energy cropping and ecosystem enhancement, as well as drawing on experiences in other jurisdictions such as the US and the EU. Differentiated support measures for bioenergy could play an important role in supporting emerging technologies and energy crops with benefits for ecosystem health in both countries. These could take the form of structured mandates for biofuels and electricity, feed-in tariffs or differentiated tax treatment for biofuels with social and/or environmental benefits. For these measures to be successful, they also need to be supported with targeted extension services, research and negative incentives to protect against threats.

As discussed at the start of the chapter, the analysis of Australia and Brazil has been based on an abbreviated policy framework and the policy measures suggested are yet to be subject to community consultation. Before the suggested measures were implemented or even developed into full-blown policy proposals, they would need to be subject to a participatory policy development process that included consideration of the goals, principles and values held by affected stakeholders. They would also inevitably be affected by political considerations and vested interests. However, whichever pathways are ultimately chosen by the governments and citizens of Australia and Brazil, the suggestions highlighted here show that it is possible to design policy measures that preferentially support energy crops that contribute to ecosystem health and socio-economic sustainability.

References

Australian National Audit Office (2015) *The Ethanol Production Grants Program*, Commonwealth of Australia, Canberra.

Azuela, G. E., Barroso, L., Khanna, A., Wang, X., Wu, Y. and Cunha, G. (2014) *Performance of Renewable Energy Auctions: Experience in Brazil, China and India*, World Bank, Washington, DC.

Barros, S. (2014) *Brazil: Biofuels Annual*, USDA Foreign Agricultural Service, Washington, DC.

Baumber, A. (2012) "Harnessing bioenergy as a driver of revegetation: an analysis of policy options for the New South Wales Central West, Australia", PhD thesis, University of New South Wales, Sydney.

Baumber, A. P., Merson, J., Ampt, P. and Diesendorf, M. (2011) "The adoption of short-rotation energy cropping as a new land use option in the New South Wales Central West", *Rural Society*, 20: 266–79.

Buckman, G. and Diesendorf, M. (2010) "Design limitations in Australian renewable electricity policies", *Energy Policy*, 38: 3365–76 (addendum 38: 7539–40).

Carbon Market Watch (2010) "Plantar – pig iron project, Brazil", http://carbonmarketwatch. org/campaigns-issues/plantar-pig-iron-project-brazil (accessed 10 July 2015).

Delta Electricity (2010) "Delta Electricity reduces carbon emissions with renewable biomass energy", www.de.com.au/Power-Stations/Wallerawang/biomass/Biomass/ default.aspx (accessed 1 September 2010).

Department of Planning and Infrastructure (2012) *Draft Strategic Regional Land Use Plan: New England North West*, Department of Planning and Infrastructure, Sydney.

Dockerty, T., Appleton, K. and Lovett, A. (2012) "Public opinion on energy crops in the landscape: considerations for the expansion of renewable energy from biomass", *Journal of Environmental Planning and Management*, 55: 1134–58.

Dovers, S. (2005) *Environment and Sustainability Policy*, Federation Press, Sydney.

Environmental Protection Agency (2013) *EPA Proposes 2014 Renewable Fuel Standards, 2015 Biomass-Based Diesel Volume*, United States Environmental Protection Agency, Washington, DC.

Gibbs, H. K., Rausch, L., Munger, J., Schelly, I., Morton, D. C., Noojipady, P., Soares-Filho, B., Barreto, P., Micol, L. and Walker, N. F. (2015) "Brazil's Soy Moratorium", *Science*, 23 January: 377–8.

Lima, M. B. (2012) *An Institutional Analysis of Biofuel Policies and their Social Implications Lessons from Brazil, India and Indonesia*, United Nations Research Institute for Social Development, Geneva.

NSW Farmers' Association (2012) *Submission to the NSW Government in Response to Delivery of the Strategic Regional Land Use Policy*, NSW Farmers' Association, Sydney.

Office of the Renewable Energy Regulator (2012) "REC Registry", www.rec-registry.gov.au (accessed 13 March 2010).

Porras, I., Barton, D. N., Miranda, M. and Chacón-Cascante, A. (2013) *Learning from 20 Years of Payments for Ecosystem Services in Costa Rica*, International Institute for Environment and Development, London.

REN21 (2015) *Renewables 2015: Global Status Report*, REN21, Paris.

Romijn, H., Heijnen, S., Colthoff, J. R., Jong, B. d. and Eijck, J. v. (2014) "Economic and Social Sustainability Performance of Jatropha Projects: Results from Field Surveys in Mozambique, Tanzania and Mali", *Sustainability*, 6: 6203–35.

Solar Choice (2011) "Solar shock: government announces retrospective NSW solar bonus scheme cuts", www.solarchoice.net.au/blog/nsw-solar-bonus-scheme-shock-government-announces-feed-in-tariff-reductions (accessed 13 November 2011).

Stephen, J. D., Mabee, W. E. and Saddler, J. N. (2013) "Lignocellulosic ethanol production from woody biomass: the impact of facility siting on competitiveness", *Energy Policy*, 59: 329–40.

Tal, A. (2015) "The implications of environmental trading mechanisms on a future Zero Net Land Degradation protocol", *Journal of Arid Environments*, 112, Part A: 25–32.

Treasury (2011) *Strong Growth, Low Pollution: Modelling a Carbon Price*, Commonwealth of Australia, Canberra.

United Nations Conference on Environment and Development (1992) *Rio Declaration on Environment and Development*, United Nations Conference on Environment and Development, Rio de Janeiro.

URS Australia (2009) *Oil mallee industry development plan for Western Australia*, URS Australia Pty Ltd for Oil Mallee Association of Western Australia Inc (OMA) and the Forest Products Commission (FPC), Perth.

Verve Energy (2006) "Integrated wood processing", www.verveenergy.com.au/mainContent/sustainableEnergy/OurPortfolio/iwp.html (accessed 20 August 2011).

Chapter 10

Conclusion

The introductory paragraphs of this book began with a kind of "doomsday" scenario for energy crops, in which the world's remaining forests were under peril, greenhouse gas emissions would soar and vulnerable people would be forced off their land. Hopefully, the real-world examples, emerging technologies and policy options presented in this book have demonstrated that a different future is possible for energy cropping. We must remain vigilant in protecting important environmental and social values against the threat of unsustainable forms of energy cropping. However, there is also great potential to deliver a future in which energy crops play an important role in meeting the world's energy needs, combating climate change, providing livelihood options for local communities and restoring ecosystem functions on degraded land.

The first chapter of this book looked at what sustainability means in the context of energy crops and set an aim of promoting a broad understanding of this concept. This broad understanding requires a focus not only on mitigating the risks that unsustainable energy crops might pose, but also involves the promotion of positive, multifunctional outcomes. Willow and poplar crops in Europe are the clearest examples of perennial energy crops delivering these multifunctional outcomes at present, including habitat for biodiversity, increases in soil organic matter, carbon sequestration and phytoremediation of contaminated soils. However, other examples cited in Chapter 4 also show that this pathway is possible for energy crops in other parts of the world, including eucalypts in Australia and Brazil, switchgrass in the US, and even first-generation crops such as sugarcane, oil palm and jatropha, where used wisely.

Much uncertainty persists around whether ecosystem health benefits can be reliably delivered through bioenergy cropping systems, as well as whether such systems will be economically competitive with other land uses and energy supply options. This uncertainty may lead some to doubt that sustainable energy crops of this nature will ever play a significant role in meeting the world's growing demand for energy, much less in the restoration of the worlds' degraded and vulnerable land. Certainly, the economic analysis presented in Chapter 7 suggests that, apart from limited cases such as willow crops in Europe, perennial energy crops with restoration benefits can struggle to compete with first-generation biofuel crops, bioenergy from wastes and fossil fuel-based energy options. This is likely to remain the case unless there are significant changes in technological, economic and policy conditions. However, the good news is that these conditions are indeed changing and the policy decisions we make today can help to determine the role that energy crops will play in the future.

Bioenergy technologies are advancing in key areas such as second-generation biofuels and combined heat and power. New knowledge and production efficiencies continue to emerge around energy cropping systems, such as oil mallee in Australia, switchgrass in the US and multifunctional agroforestry systems involving tree crops, oilseeds and cattle grazing in Brazil. As the effects of climate change increase, concerns around habitat connectivity and soil protection are likely to grow, along with efforts to reduce or offset greenhouse gas emissions. All of these trends point to a future in which bioenergy, agroforestry and ecological restoration will increasingly come together.

While some global conditions may shift in a way that favours a greater role for sustainable energy cropping systems in the future, other trends may work against this outcome. Some forms of energy cropping continue to be associated with negative environmental and social impacts, particularly oil palm in southeast Asia, soy in Latin America and jatropha and other emerging crops in Africa. These examples of deforestation, dispossession and degradation discussed in Chapters 3, 5 and 6 threaten biodiversity, soil and water quality, food security and livelihoods for local people. Moreover, they also damage the reputation of energy crops across the globe and create the risk of consumer boycotts or one-size-fits-all policy measures that restrict the development of more sustainable forms of energy cropping.

Further work is needed to protect vulnerable communities, forests and other high conservation value land from the spread of unsustainable cropping systems, whether they be for energy, food or any other purpose. Fortunately, there are examples we can draw on where environmental destruction has been reversed or slowed down, such as the 70 per cent drop in Brazil's deforestation rates cited in Chapter 3, which was achieved through a combination of international pressure, intervention by key participants in the soy and beef supply chains, government reforms to the implementation of the Forest Code and behavioural change among landholders (Nepstad et al., 2014).

Further research is required to address the threat of unsustainable energy crops, as well as the barriers to more sustainable options. This includes both research into the impacts of energy cropping as well as the effectiveness of different policy interventions. This should include environmental analysis to understand the impacts of different forms of energy cropping on biodiversity and soils, social analysis to understand how landholders and local communities can benefit rather than suffer from the introduction of new cropping systems, and policy analysis to understand how positive and negative incentives can be combined with other measures, including extension, technology change and institutional change.

Having set out a broad conceptualisation of energy crop sustainability in this book that considers both threats and potential benefits, we now turn to the question of how such a conceptualisation can be promoted among policy-makers, NGOs and other stakeholders. This requires returning briefly to the definitions of sustainability and sustainable development discussed in Chapter 1.

Sustainable development requires that human welfare and social equity are enhanced by activities that are economically viable and ensure the long-term persistence of key resources and environmental values for future generations. Looking at the criteria cited in Chapter 1 from the United Nations (UN), the European Union (EU) and the Roundtable on Sustainable Biomaterials (RSB), more specific expectations emerge when it comes to bioenergy. The production of energy crops should not destroy forests or other high value ecosystems, nor should it take land away from vulnerable people, increase levels of food insecurity or exploit workers. In relation to greenhouse gas emissions, energy crops are expected to not simply maintain the status quo, but to play a positive role in reducing the emissions-intensity of our fossil fuel-based economy. Similarly, when it comes to socio-economic development, energy crops should be an active part of the solution rather than simply maintaining the status quo.

On top of these existing expectations, this book has sought to add the expectation that energy crops can and will play a role in restoring degraded land by enhancing soil health, improving water quality and providing habitat for biodiversity. This is a view of energy crop sustainability that is not strongly reflected in the biofuel standards from the EU and RSB, or in the standards of the Forest Stewardship Council (FSC) and Sustainable Agriculture Network (SAN), which have also been used as key references throughout this book. By and large, these standards consider issues of biodiversity, water quality and soil health from the perspective that energy cropping (and cropping in general) has the potential to pose threats that must be mitigated.

Changing the prevailing perspective in international sustainability standards involves shifting one's thinking further into the overlap zone between energy cropping and ecological restoration. Chapter 4 highlighted how the concepts of conservation through sustainable use and multifunctionality can help to conceptualise sustainable development in a way that brings together energy cropping and ecological restoration. As discussed in Chapter 4, the principles of conservation through sustainable use can be applied not only to "natural" or "wild" ecosystems, but also to cases where vegetation is being established for a combination of commercial and environmental reasons.

Modifying our conceptualisation of energy crop sustainability to include active enhancement of ecosystem functions and services does not necessarily mean that all energy crops must enhance conditions to be considered "sustainable". Nor does it mean that energy crops that simply maintain rather than enhance ecosystem functions should be considered "unsustainable". Rather, we should follow the idea promoted by Stephen Dovers and others that sustainability is a general direction for society rather than a clearly measurable endpoint (Dovers, 2005). Following this idea, forms of energy cropping that *maintain* biodiversity and soil health help us move some way towards sustainability (especially in a world where many ecosystems are moving in the other direction), but energy cropping systems that actively *enhance* biodiversity, soils and other ecosystem values move us even further in the direction of sustainability. For policy-makers and those working in the bioenergy, agriculture and restoration sectors, the goal should be to create the conditions under which the most sustainable forms of energy cropping can thrive and to find ways to promote energy crops that have benefits for ecosystem health over those that do not.

The approach that has been taken towards climate change mitigation within the bioenergy sector highlights one pathway to integrate an expectation of environmental enhancement into notions of bioenergy sustainability. As shown in Chapter 2, different forms of bioenergy can have very different life-cycle emissions. First-generation biofuels tend to perform worse than advanced biofuels or solid biomass fuels used for electricity and heat. This is partly due to their greater likelihood of contributing to forest-clearing (either directly or indirectly), as well as factors such as fertiliser use and inefficient conversion technologies. In some cases, the life-cycle emissions of biofuels may be much worse than using fossil fuels, such as where tropical peat forests are cleared to produce palm oil for biodiesel (Gibbs et al., 2008). At the other extreme, some energy cropping systems may not only reduce emissions compared to fossil fuels, but may actually draw carbon dioxide out of the atmosphere, producing *carbon-negative* biofuels (Mathews, 2008).

The differing potential impacts of biofuels in regards to greenhouse gas emissions has given global policy-makers and industry stakeholders a chance to think about what our expectations of biofuels should be in different contexts. Should we accept biofuels as "sustainable" if their life-cycle emissions are worse than fossil fuels? Should we expect all biofuels to achieve a certain level of GHG saving relative to fossil fuels (e.g. 35%, 50% or 60%)? Should we expect all biofuels to be carbon-negative?

While the answers to these questions may vary over time and between different stakeholders, certain trends have become apparent. When the Kyoto Protocol was drafted in the late 1990s, it was acceptable to regard all biofuels as carbon-neutral and to encourage their use equally. By the time that that the EU adopted sustainability criteria under its Renewable Energy Directive in 2009 and the US revised its Renewable Fuel Standard in 2010, much more was known about the emissions from deforestation and fertiliser use associated with some biofuels. Both institutions introduced GHG benchmarks that restricted their support programmes to fuels with demonstrated GHG savings relative to fossil fuels. The RSB has gone further than this, using GHG benchmarks not just to determine whether a biofuel is deserving of government support, but whether it is can be regarded as "sustainable" at all.

Our future expectations around energy crops and ecological restoration could follow a similar pathway to that which has been followed in relation to biofuels and GHG emissions. For example, energy crop producers may need to show they are at least maintaining conditions (i.e. not degrading habitat or soils) in order to meet a basic standard of sustainable production (e.g. the RSB standard). Energy crops that go beyond this and actively enhance conditions for soils, water quality or biodiversity would become eligible for policy support programmes (e.g. mandates or tax breaks), with the level of support increasing in accordance with the level of benefit. This approach would recognise that such energy crops are more than simply "not unsustainable", but actually help to move the world in the direction of sustainability.

The approach described above was employed for the two policy case studies presented in Chapter 9. These case studies showed how it possible to structure policy mechanisms such as mandates, tax breaks and feed-in tariffs to preference forms of energy cropping that not only produce renewable energy, but also provide environmental and/or social benefits through the production system used. The incorporation of local environmental and social co-benefits into renewable energy incentive schemes is an emerging opportunity that is yet to be fully capitalised on. Isolated examples exist of where this has been done, such as the promotion of "social fuel" under Brazil's biodiesel programme (Chapter 6) or landscape preservation under Germany's feed-in tariffs (Chapter 8), but there is significant potential for such an approach to be employed in bioenergy support schemes across the globe.

Chapter 9 showed how local environmental and social co-benefits could be incorporated into schemes such as Australia's Renewable Energy Target, as well as how Brazil's use of the "social fuel" label could be expanded to promote "ecosystem fuel" as well. These changes would require rigorous verification systems, increased knowledge around the impacts of energy cropping and complementary measures to protect against risks of environmental damage. However, such changes would also serve to refocus the incentives that these schemes provide for renewable energy production towards cropping systems that also provide landscape-scale environmental enhancements.

There are significant challenges ahead in moving towards a future in which energy crops play an increasing role in both supplying clean energy and protecting and restoring degraded ecosystems. Continued research and development support is required to enhance the viability of new cropping and processing systems. Protections for vulnerable ecosystems and vulnerable people need to be maintained and strengthened. An adaptive management approach is required to effectively design and implement new policy measures that can respond to uncertainty and adjust to new information. However, if these challenges can be effectively managed, the importance of energy cropping as a form of energy production, socio-economic development and environmental enhancement can grow well beyond its current levels. Along the way, these changes may also help to redefine what we think of as sustainable energy in the future.

References

Dovers, S. 2005. *Environment and Sustainability Policy*, Federation Press, Sydney.

Gibbs, H. K., Johnston, M., Foley, J. A., Holloway, T., ChadMonfreda, Ramankutty, N. and Zaks, D. (2008) "Carbon payback times for crop-based biofuel expansion in the tropics: the effects of changing yield and technology", *Environmental Research Letters*, 3: 1–10.

Mathews, J. (2008) "Carbon-negative biofuels", *Energy Policy*, 36: 940–45.

Nepstad, D., McGrath, D., Stickler, C., Alencar, A., Azevedo, A., Swette, B., Bezerra, T., DiGiano, M., Shimada, J., Ronaldo Seroa da Motta, Armijo, E., Castello, L., Brando, P., Hansen, M. C., McGrath-Horn, M., Carvalho, O. and Hess, L. (2014) "Slowing Amazon deforestation through public policy and interventions in beef and soy supply chains", *Science*, 6 June: 1118–23.

Index